# Conversations about Challenges in Computing

Are Magnus Bruaset · Aslak Tveito
(Editors)

# Conversations about Challenges in Computing

 Springer

[ simula . research laboratory ]
- by thinking constantly about it

**Editors**
Are Magnus Bruaset
Aslak Tveito
Simula Research Laboratory
Fornebu, Norway

ISBN 978-3-319-37539-7     ISBN  978-3-319-00209-5 (eBook)
DOI 10.1007/978-3-319-00209-5
Springer Cham Heidelberg Dordrecht London New York


Math.Subj.Classification (2010): 65-XX, 68-XX

Printed on acid-free paper


Springer is part of Springer Science+Business Media (www.springer.com)

# Preface

We go to work every day, picking up our projects where we left them the day before. It's a mess, of course: complex models, software densely populated with bugs, dubious input data, and so forth.

Most of our attempts to understand matters fail, but every once in while we see the light and are able to add a small contribution to our common collection of scientific knowledge. Enormous amounts of time, money and energy are invested in these attempts to comprehend the world around us. In this light, we should pause now and then to consider which problems most deserve our attention.

To celebrate the tenth anniversary of Simula Research Laboratory, we invited outstanding scientists from around the world to present their recent achievements and their view on future challenges. The talks were intended to fuel vivid discussion at Simula concerning where our scientific efforts are called for. Given this unique assemblage of scientific leaders in communication technology, software engineering, scientific computing, and computational science, we also invited two celebrated writers to interview them. The results of these conversations are presented in this text, which we hope you will enjoy. In addition you will video clips of all talks from the anniversary conference on the special web site http://challenges.simula.no.

If you are curious to learn more about Simula, we encourage you to browse our web site, www.simula.no, and have a look at the factsheet "This Is Simula" on page 101.

Fornebu, February 2013

Professor Are Magnus Bruaset
*Director of Research at Simula Research Laboratory*

Professor Aslak Tveito
*Managing Director of Simula Research Laboratory*

# Challenges in Computing
# December 14–15, 2011

All talks were filmed and are available at challenges.simula.no

**Conference program, day 1**

Welcome and introduction
*Professor Aslak Tveito, Managing Director of Simula Research Laboratory*

Opening and announcement of the winner of the Computational Science and Engineering Prize 2011[*]
*Tora Aasland, Norwegian Minister of Research and Education*

High-resolution simulation of mantle flow and plate tectonics
*Professor Carsten Burstedde, University of Bonn*

US Ignite
*Professor Keith Marzullo, University of California San Diego & Division Director at the National Science Foundation*

Engineering Software in the Future Era: The Role of Uncertainty
*Professor Paola Inverardi, University of L'Aquila*

Grand Challenges in Computational Inverse Problems with Illustrations from Geophysics
*Professor Omar Ghattas, University of Texas at Austin*

**Conference program, day 2**

The Challenges of Computer-based Prediction
*Professor Martin Shepperd, Brunel University*

Simulating Cardiac Function and Dysfunction
*Professor Natalia Trayanova, Johns Hopkins University*

Requirements for Pervasive Privacy
*Professor Bashar Nuseibeh, The Open University & Lero – the Irish Software Engineering Centre*

Model Reduction, Complexity Reduction
*Professor Alfio Quarteroni, École Polytechnique Fédérale de Lausanne, and Politecnico di Milano*

Challenges in the Evolution of the Internet of Things – Processing Real-World Information in the Cloud
*Dr. Heinrich Stüttgen, Vice President of NEC Laboratories Europe*

---

[*]  The CSE Prize is awarded by Springer-Verlag. The 2011 prize was awarded to Laura Alisic, Carsten Burstedde, and Georg Stadler
for their outstanding work on simulating global mantle convection at tectonic plate boundary-resolving scales.

# Contents

# The Nature of the Beast

*An Interview with Olav Lysne by Kathrine Aspaas*

Professor Olav Lysne is on vacation – which means living without a computer network connection at his holiday cottage outside Fredrikstad in southeast Norway. Because a computer network is work. Not only in the shape of emails, online news sites and other digital temptations – we are talking about the *network* itself – the myriad of mainframes, masts, copper wires and routers. That is the area of expertise to which the professor is devoted. For what do we really know about the network's architecture? A lot less than we think. We have created a beast, whose full extent we cannot see and which we have made ourselves dependent on. And this is precisely where Professor Lysne's research project begins. How does this beast behave? And – not least – how can we make network access more robust, given the beast's many unpredictable quirks and caprices? It is certainly no easy matter to obtain a complete overview of the beast's internal organs, for they are governed in large measure by politics and trade secrets.

"Telenor knows its bit of the network. NetCom knows its bit. Ice knows its. They cooperate through agreements which are business critical and therefore confidential. They keep the network's structure secret, partly for commercial reasons, but just as much for reasons of security. Because there are people out there with malicious intent, and there are good reasons to keep the infrastructure undisclosed. So we must accept the unpleasant truth that the network is, and will probably remain, relatively closed and extremely difficult to pin down. We just have to make the best of the situation."

## Robust Networks

Let us take a little journey back in time. Around 30 years. To when the internet as we know it today did not exist. Initially, in the early 1990s, it was seen as a bit of fun. Then we got emails and mobile phones on a wide scale, and around 1995 it switched from being fun to being useful. In the years to 2000 it went from being useful to important, and since then that importance has grown and grown. At the same time, a new entertainment perspective has emerged, where games, television and music now rule the roost. And this is where we find the professor's concern.

*'There is no reason for any individual to have a computer at home.'*
Ken Olsen, Chairman of the US Digital Equipment Corporation (1977).

"Today it is perfectly valid for an operator to say that it doesn't matter if the network is down for a few hours or a couple of days, as happened in Norway during the storm Dagmar. This means more or less that you cannot update your Facebook status for a few hours. It contains a sufficiently large pinch of truth that they get away with saying it now. But in the coming years the network will transition from being important to being indispensable, and this is where we bump up against major social issues. How are we going to tackle the coming sharp rise in the number of elderly people, for example?"

A. Bruaset, A. Tveito (Eds.), *Conversations about Challenges in Computing*,
DOI 10.1007/978-3-319-00209-5_1, © Springer International Publishing Switzerland 2013

▶    You do realise that those elderly people are you and me?

"Yes, I'm only too aware of that. Why do you think I'm working on this? Pure self-interest! And how are we going to meet our need for care? That's right, we will be allowed to live at home for as long as possible. But I'm not entirely comfortable with the prospect of the network being down for a few hours or a couple of days if my pacemaker is connected up to it. And if we want to save energy by having smart cars and smart houses, all that will be network based. We are on our way there now, because we have already seen examples where computer networks have halted water supplies. They have prevented doctors from talking to each other and grounded aircraft. They prevent the police from working together. That takes us back to the summer of 2007, when a fire at Oslo's main train station put the capital's anti-terror capability out of action for almost 24 hours. Because of data communication problems. What I'm trying to say with all this is that data communication has now become a *single infrastructure of failure* – in other words, that all other infrastructures will lie on top. Hospitals. Air traffic. Police. Almost all the infrastructures we have to sustain society will have the internet as their cornerstone. Which means we need a more robust network."

At this point we have completely bought into Lysne's message: computer networks are extremely important, and they are merely getting more and more important. We must accept that we are not going to get much insight into the network's architecture, for commercial and political reasons. So now we should be ready for the professor's research agenda.

"Until recently we have been preoccupied with solving technical problems, on the assumption that we have a complete overview of how the system is built up. Now we are being forced to work in a different way – on the assumption that we *don't* know how the system is built up. And there are two things we must do:

1)  Study this in the same way we study nature. By observing how the great network beast behaves, and on that basis try and understand as much as possible about how it is constructed.

2)  When we have understood how it behaves, we can, hopefully, use this knowledge to create applications that are robust in the face of anything that might happen. How will my pacemaker behave if a power line falls down somewhere? How can we ensure it keeps on working anyway?"

▶    But what about openness – the possibility of making a better map. Won't we someday dare to make it open?

"No, I don't think we would ever dare to do that. We will keep organisations like the Norwegian Post and Telecommunications Authority (NPTA), which is both a watchdog and a collector of information. We may envisage a situation in which they insist on having a complete overview themselves, but keep that information as close to their chest as the network operators do. And with good reason. We may also envisage them sharing information with selected, critical entities that they trust. But we're not there yet."

▶    So we have to keep studying the beast from the outside. How do you do that in practice?

"For example, by measuring the networks' uptime. We carried out a major measuring exercise in the run-up to the electronic voting system trial in the autumn of 2011, which included ten local authorities. The Ministry of Local Government and Regional Affairs wanted to test how certain they could be that the networks would be up during the election trial. Since the NPTA did not provide that information, they commissioned us to measure uptime over an entire year – for all operators – in all polling stations. We discovered, for example, that there must be some kind of undisclosed relation between Telenor and NetCom, because when Telenor has problems in its core network, NetCom's customers can also be hit far harder than Telenor's. This is just one example of the kind of analysis we perform. We are currently in the process of rolling out a new structure in which we will have 500 measuring points. It is a scheme that we've got two of the operators – Ice and Tele2 – to fund."

▶      But don't they also lease network capacity from Telenor?

"Actually Ice has most of its long-distance capability through Broadnet. In Norway there are two networks with pretty much nationwide coverage. Telenor owns one, while the other was originally built up by the Norwegian railway company NSB – called Banetele. They laid fibre-optic cables along the tracks. When we talk about the construction of new mobile networks, what we mean are new base-stations linked to one of these major long-distance networks."

▶      Won't it gradually get cheaper to build networks?

"The bulk of the expense is associated with work on the large masts, while the equipment to be installed doesn't have to be so costly. When you phone me in Fredrikstad it is easy to imagine that the signal is carried over the airwaves all the way, but it's not. It is transmitted wirelessly to the very first base-station, perhaps a kilometre or two here in Oslo – and then travels down a wire almost all the way to me, before switching back to a little wireless network for the last little bit."

▶      And where does the Cloud come into the picture?

"Now we're talking about another of my activities, which is related to this one. We are used to having everything stored locally on our PCs, but the trend now is to construct vast datacentres, so that more and more of our data will be stored there instead. The people who use GoogleDoc, for example, don't have Word installed on their machines. All their documents are hidden externally in the Cloud."

▶      Do you use it?

"No, I have all my data here on my PC."

▶      Don't you trust the Cloud?

"A professional cloud provider like Google would probably be able to take better care of our documents that we can manage ourselves. But that brings us straight back to the matter of networks. And what you have stored in the clouds will only be accessible as long as the network is working. No network, and you are in real trouble."

We're back at the beast again. And I just can't bring myself to accept being unable to create a functioning map of the network's internal organs. It seems too ineffective and too vulnerable. After all, we are talking about a critical infrastructure that should most definitely be a collective benefit.

▶      Don't we also need global observation and international management
         tools and bodies?

"Quite right. But that puts us far into the realm of politics, which is not my sphere. And we have to get this in place in each individual country first. My job is to obtain as much information as possible about how the network behaves, from the outside. And that is a paradigm shift in the field of network research."

So let's sum up the situation so far. We must accept secret agreements between operators. We must accept secrecy from government agencies. Not even Norsk Helsenett, which is creating a communications network for the emergency services and healthcare sector, or the Ministry of Local Government and Regional Affairs in connection with electronic elections, have access. These are the fundamental operating conditions. So we must learn to understand them as well as possible from the outside. We must treat the network as biologists would observe a living organism.

## Music Maker

The living organism that is Olav Lysne is described by those who know him as organised, articulate and musical. He perceives himself as that irritating schoolboy who knew all the answers in class. He gave his musicality an outlet through intense trumpet-playing in a marching band until he had children 21 years ago. Today you will find him blowing his trumpet in the horn section of the Simula Band, and he has plans to take guitar lessons from fellow professor Hans Petter Langtangen. His passion for music takes in all four points of the musical compass: pop, rock, jazz and classical.

Informatics (computer science) captured him as long ago as 1983, when he went to study at the University of Oslo, and came into contact with some of the period's rare computers. Writing programs – getting the machine to do what he wanted – such a lot of fun! He took his PhD in mathematical modelling, where the aim was to find mathematical proof that programs worked. In the mid-1990s it became clear that he wanted to go into research, and the place he found an opening was in the field of network research. Since he was now publishing articles in two different fields, he quickly gained a professorship. And he transferred over when Simula started up in 2001. He now considers the fact that he was forced over into network research because there was no room in mathematics as a blessing in disguise. It forced him to change his focus, learn new things – quite simply, become multidisciplinary. It reminds me of something I have heard wise people say, that it is on the path less travelled that the action happens. That it is off the beaten track that we find new highways and new solutions.

However, Lysne is far from being alone in studying our vulnerable and oh-so-vital network beast. As head of the *Simula Network Systems* research department he has 17 permanently employed researchers, post-doctoral fellows and PhD students on his team – most of them with a background in informatics. Half of them are engaged in the network beast project, and they are not simply making observations. The idea is also to produce prototypes – mechanisms that can tame the network beast's unwanted behaviour.

We can, for example, return to the growing numbers of elderly people and our pacemakers, which – at some point in the future – will be connected to the internet. Lysne and his team could make a prototype which ensures that our pacemaker works, even when the network fails.

▶    What is that kind of prototype actually like?

"Let's say that we are observing an area around Sandnes. We see that there is no connection between when NetCom's network goes down and when Telenor's network goes down. We can use that information to ensure that our pacemaker is connected to both networks. Simply put, we can make a box that is connected to both networks and which communicates alternately between them both. Then we can measure how, with this functionality enabled, it would handle any problems with the network. For us a prototype is something which is inspired by a finished product. Our job is to find out how a product of that kind could work, and publish the result. We do all this to answer questions, and our output is insight."

▶    Where does your funding come from?

"We receive a little state funding from the Ministry of Education and Research, but most of the money comes directly from our major projects. We have a substantial project called *Robust Networks*, to which the Ministry of Transport and Communication has contributed NOK 5 million this year. In 2013 that figure will increase to NOK 7 million. The project started up in 2006. We also receive quite a bit of money from the Research Council of Norway, and some from the EU's Seventh Framework Programme (FP7). In addition our industrial partners contribute some funding. We get a bit from Tele2 and a bit from Ice. On the Cloud side we have a close collaborative relationship with Oracle – formerly Sun."

▶   Why aren't the big players – Telenor and NetCom – involved?

"That's a good question. Our impression is that they are not terribly interested in anyone rocking the boat. But we are constantly striving to bring everybody on side, and build up the players' belief in us over time."

▶   Do you collaborate with other research environments?

"Both the University of Oslo and the Norwegian University of Science and Technology (NTNU) are on board. We also collaborate with *Uninett*, which is the network operator for universities and colleges in Norway. They are extremely good and are open about their network structure, as is *Unik*, which is the university campus at Kjeller. We also work closely with a British, a Spanish and three American universities. The University of San Diego in particular has been performing measurements for a good while."

▶   What kind of timeframe do you have?

"We reckon on a decade. And the very first thing we must do is find out how we should attack this beast. We have to develop mathematical and statistical methods – put together existing methods in a way that is appropriate for what we are going to study. Give me three or four years and I will be able to say more about our subsequent timeframe."

▶   This beast… it has a physical form, made of copper, fibre, base-stations and masts. But does it also have a mind?

"That's a good point… We have operator agreements and management structures. Factors that are difficult to get to grips with, like corporate operating cultures and routines. Not to mention the trust between operators. We have learned a lot about all this in the past year, as we have talked to the various companies. Trust is currently almost non-existent, and we have already identified an invariant: the bigger the company, the less they are trusted by those around them. This may be a shared human phenomenon, that big things are a bit scary. In addition to a physical form and a psyche, the network beast also has a kind of free will, for the various players will pull in different directions. Some suppliers, like *Uninett*, have a public-service mission. They open their network for research. While commercial players have a bottom-line mission, and we are already coming across mechanisms which we understand are for accountants and economists, and which surprise us. Why won't Telenor and NetCom join our research project? That kind of things. Then we start to understand that we may need a financial specialist on the team. We are trying to analyse a beast which, because we are measuring and examining it, is starting to behave differently. That is also part of the picture."

▶   And every country has its own beast that they are trying to understand and take care of?

"Yes, at the end of the day all these networks are connected. But we must first learn to crawl before we can walk. We do that initially by understanding the situation in Norway – a small, western country, with a suitably complex operator situation, and a varied geography and population."

▶   And phase two?

"Now you are asking about something far down the road. I dream of a time when researchers arrive at the point where they have accessible measurement data from the situation in as many countries as possible – accessible for all. Based on that we can learn more about the impact of regulatory decisions on the robustness of the network. For example, say the NPTA issues a directive regarding the out-placement of diesel tanks at all base-stations in case there is a power outage. What will be the effect of being able to operate emergency generators? That issue was actually discussed after the storm Dagmar. A directive may also be issued ordering operator A to have an agreement with operator B so that they can back up each other's traffic. What effect would that have on robustness? We want to be able to say something quantifiable about it."

► What is needed to arrive at that situation?

"We must acquire a sound and well-thought-out grip on the technological solutions, operators' responsibilities, regulatory provisions and their impact. On top of everything else, we need a level of cooperation between the various players that provides us with greater predictability and stability. So we can actually arrive at the point where we have my blood-pressure monitor and pacemaker online, without me having to go around worrying that the network might go down. It is an expressed political ambition that we – us elderly – should be able to live at home for as long as possible. A precondition for that is access to technologies that provide robust monitoring of, for example, heart-starters and pacemakers."

## Machines, Humans and the Beast

It's perfectly possible to project a lot of humour into a future like that – a kind of Orwellian surveillance society that has run completely amok. Where our hearts (and why not our brains, too?) are monitored (and why not controlled?) by large data centres in Bangalore, India. More realistically, however, it's possible to imagine a more trusting and holistic scenario, where we must try and understand the coexistence between humans, society and machines.

► What do you mean by holistic in this context?

"In the field of informatics the whole is typically made up of a web of sub-specialisations, with individuals being experts in one or two of them. But the really big challenges now lie in finding the connection between all the sub-specialisations. There is not much to be gained by working solely on robustness, as we have up to now. I would argue strongly that we need a multidisciplinary approach within our own field."

► What is standing in the way of such a multidisciplinary approach?

"It is difficult to find specialists within all the fields concerned. Another challenge lies in differences in cultures and ways of speaking, and different places we publish articles. On top of which it's not exactly career-enhancing. If I start collaborating with people who are specialists in a field that is different from mine, it will take a year or two before it begins to bear fruit. It is therefore a long-term investment that happens far too rarely. It is more tempting to continue pushing out results in an area we are already familiar with."

► That is reminiscent of the lament articulated by my new hero, the mathematician Richard Hamming, who said that we are currently working while standing on each other's toes… in a kind of misguided completion.

"Yes, that's a good one. I recognise the situation. I can recall that when we started Simula almost a decade ago, our objective was to create a research environment that we imagined Richard Hamming would want to work in."

► So what do you do at Simula to step off each other's toes and work in a multidisciplinary way?

"The various universities and partners that we have, have been selected to cover sub-specialisations. Nevertheless, we must be honest enough to say that the multidisciplinary aspect has not been prioritised."

▶     With what consequences? It has obviously not been important enough for you?

"No other consequences than that we are not as well positioned to work in a multidisciplinary way within the field. And I will work to enable us to gradually start being so positioned. You could say that up to now we have chosen to have a sub-disciplinary strategy, and now I think we should have another one."

▶     If you were to look at this in a ten-year perspective – as far ahead as Simula has existed so far – what would you see?

"A picture that is very interesting to look at is a phenomenon like Twitter. Ten years ago it wasn't even on the horizon. During a fantastically short space of time a phenomenon like that has been conceived as an idea, developed, launched, been embraced by hundreds of millions of people around the world, and played a key role in toppling regimes in North Africa. All that has happened in just a handful of years. For me that is an incredibly powerful illustration of the potency of the force for change that IT still represents. I believe that we are still right at the start of these changes. The pace is going to get faster and faster, despite the fact that we have developed a kind of fatigue with regards to the claim that IT will change everything. Our observation is, yes, it does. And it will go on doing so at an increasingly rapid rate. We have, quite simply, invented the wheel."

▶     What is it that is slowing us down?

"The fact that we've been saying 'ICT changes everything' for almost three decades. We've said it for so long that the pace of change has become part of our everyday lives – a habit…"

▶     So, it is now it really all starts?

"Well, no one would have predicted four years ago that there would be an internet phenomenon that would play a key role in toppling the regime in Egypt."

▶     I am trying to fish for an answer here… did you notice that?

"Yes, but what is the question you want an answer to?"

▶     What is it that is slowing us down? Take the plans to digitise the public administration, for example. It's a huge job, and we know it has to be done. But the main challenges are no longer technical – they are human and psychological. What happens now? Am I going to lose my job? Will I lose my desk? My office? Who will eat me now? No, I don't want to!

"Yes, I agree."

▶     This is something of a hobby-horse of mine, but you don't have to come along for the ride if you'd rather not…

"I am fairly certain you are onto something there. And I am also certain that my distinguished colleague Magne Jørgensen has a lot to say on the matter."

▶     That's right…

"But I can say something about the psychology at a dinner party. Anyone who starts talking about the digitisation of the public services in that setting will be met by a long-drawn-out yawn. For everyone takes it for granted. Ten or fifteen years ago you could mention digitisation and everyone would sit up in anticipation that something exciting was going to be said."

▶    That's rather funny, because the psychologists report the opposite. When, 10 or 12 years ago, the woman sitting next to them at a dinner party found out what they did  she would pull back in alarm. Nowadays it sparks an hour-long discussion about brain research and neuroplasticity.

"Perhaps the relaxed attitude to digitisation is attributable to the fact that the technological pace of change is something we have a solid grip on. For a while, for example, people were afraid that digitisation would create what was called the two-thirds society, one in which a third of people would be completely marginalised because they couldn't use a computer. That didn't happen. Despite all our feelings of vulnerability, we feel more comfortable with digitisation now."

The hobby-horse goes back into its stable, and we attempt once more to look forward. For example by quoting Sir William Preece, engineer-in-chief of Britain's Royal Mail, who in 1876 uttered the infamous comment: 'The Americans have need of the telephone, but we do not. We have plenty of messenger boys.'

▶    In which areas today do you consider our thinking to be as near-sighted as Preece's?

"This quote underlines primarily how careful we must be not to speak bombastically about the future. There is very little reason to assume that we are handling this any better than Preece and his cohorts did. What may moderate this statement somewhat is, once again, the matter of the pace of change. We have had a few decades of extremely rapid change, so change is something we consider to be an everyday occurrence in a way that they did not 150 years ago. But what kind of change we will see – which way it will go – is very difficult to predict."

▶    That gets me thinking about the nature of humility, and the starting point for your research project. What do we know about the networks' architecture? Almost nothing! That is why we have to conduct our research in a different way. We must approach this network beast with an open mind and ask questions.

"Yes, and now a new question has just popped into my head. Where will the beast take us? It will take us to a completely different place than where we are today – to completely different societies. What those societies will look like is difficult to predict. Take the phenomenon Avaaz.org, an activists' website that in just a few years has grown to 15 million members in 194 countries. Potentially they could take us somewhere else. Counterforces will emerge, but perhaps Avaaz – which means voice in Farsi – will develop into a major, global force for democratisation. Or perhaps it will be smothered by opposing forces. That's how developments go."

▶    We are allowed to hope …

"Yes, I have just read Steven Pinker's book about why global trends with respect to violence are actually falling. He explains that the emergence of printing and literacy put people in situations where they immersed themselves in other people's perspectives. In this way the circle of people we felt close to and cared for was widened. He draws the analogy right out to the internet and digital communication. That reading things from other people's perspectives means that we expand the circle of creatures we feel responsible for. We can see that online phenomena like Avaaz and Twitter have a similar effect. Which does not necessarily have to do with forms of government, but operates on a more psychological and cultural axis based on human compassion."

The journalist's hobby-horse was given another little outing there, and I venture to believe that the professor's interest in psychology and communication carries within it the germ for turning the intention to make Simula adopt a multidisciplinary approach into more than just a pipe-dream.

► **What is the bravest thing you have done as a researcher? When were the stakes highest?**

"Goodness! Highest stakes, hmm [*ponders at length*]. It must be in connection with what I'm doing at the moment. The switch I am making in the field of network research. Studying the network as a natural science. Finding robustness through measurement of the beast. Recognising that we currently know very little about this creature, but that we are now going to find out. We are changing from a research activity that is safe and familiar to doing something different – finding a method that has not been used before. Here lies a clear risk. And it may be that we come out in five years' time having found nothing at all. But I'm now a 50-year-old professor. If I don't do it, who will?"

► **Why is it so important that someone does it?**

"Then we are back at the start. Networks are becoming more and more important, at the same time as they are becoming less and less transparent. This is something we have to address through new methodologies."

► **What will be the biggest difficulty?**

"There are two major challenges. Getting everyone to work together, and handling the political dimension. The amounts of data will also be huge, and it will be a formidable task to draw insight from them. We have to create good programs that can handle the data, and we must guess correctly with regard to what we look for. Find the patterns in the data. How does the beast behave? We risk a situation in which we fail to find a lot that is observable because we're not searching in the right way. But it is an incredibly exciting project. Around nine people will be working on this for five years."

► **Let us fast-forward to the year 2050. How far have we come then – if you could decide?**

"I don't think I want to decide. This future will be created as the sum of so many creative brains and ideas. And what we end up with will probably be more fantastic than anything I am capable of dreaming up."

So speaks a genuinely open professorial mind, one who acknowledges that the whole is greater than any of us can conceive. May he fulfil his dream – and tame the Beast.

# Igniting the New Internet

## An Interview with Keith Marzullo by Dana Mackenzie

In 2011, Google announced its plan to bring optical fibre communications to homes in Kansas City. The optical fibre will allow residents to surf the Internet at speeds up to a gigabit per second, roughly 100 times the national average in the US. That would be enough to download a movie in a second. For people who love their online video, Kansas City might become the best place to live in the country.

Meanwhile, EPB Fiber Optics, using economic stimulus funds from the US Department of Commerce, has brought gigabit Internet to several thousand homes in Chattanooga, Tennessee. Similar projects are on tap for Cleveland, Philadelphia, and Salt Lake City. The gigabit Internet is coming, a city at a time. But what will people use it for?

At present, says Keith Marzullo of the National Science Foundation (NSF), the development of the gigabit Internet is stymied by a catch-22: People won't ask for fibre to the home if there are no services available for it and developers won't offer services if they don't see a market. At the Challenges in Computing conference, Marzullo described a new NSF initiative called US Ignite, which is intended to break this deadlock.

'Our goal is to create a national innovation ecosystem', he said. US Ignite will comprise three steps: first, to 'stitch together' the islands of high-bandwidth connectivity; second, to award grants to universities and industry to develop novel gigabit applications and services; and, third, to develop public–private partnerships to turn those applications into business plans. The project will include open competitions run in conjunction with the Mozilla Foundation. This is a new working model for the NSF, which traditionally disburses money through peer-reviewed grants. Through open, public competitions, the NSF hopes to attract ideas from innovators who might be cut out of the loop in the regular funding process. For example, the next great ideas may come from high school or college students or a college dropout such as Bill Gates or Steven Jobs.

On the hardware side, US Ignite will be facilitated by an already existing NSF-funded infrastructure called Global Environment for Network Innovations (GENI). GENI makes it possible to carve off slices of the Internet to experiment on without affecting the rest of the Internet; thus, innovators can test their ideas, on a reasonably large scale, in places where gigabit access already exists.

US Ignite finally had its formal launch at the White House on June 14, 2012. Marzullo attributes the delay to the search for high-profile leaders from the private sector who can give the project visibility and credibility.

In the interview below, Marzullo talks about the plans for US Ignite and also describes his experience as a successful researcher crossing over into the less familiar territory of science policy. The interview was conducted on February 25, 2012.

A. Bruaset, A. Tveito (Eds.), *Conversations about Challenges in Computing*,
DOI 10.1007/978-3-319-00209-5_2, © Springer International Publishing Switzerland 2013

> ► Your biography says that you got your PhD at Stanford University in 1984. What were your research interests then?

When I started at Stanford, I was actually in applied physics, studying force-free magnetic fields on the sun. But I decided that there was no career in that and switched over to computer science, although I was technically in the electrical engineering department. The research I did for my PhD grew out of my work as a research intern at Xerox. I developed one of the first clock synchronization algorithms, which was deployed on the Xerox Research internet and later patented by Digital. Part of it still lives on. If you Google *Marzullo's algorithm*, you'll find an algorithm that grew out of my thesis.

> ► Why is it important to have clocks synchronized on the Internet?

If you have a common time base, it allows you to do things such as debugging. As you're collecting logs, events will be time stamped and it would be nice to put them into some kind of temporal sequence, which requires synchronized clocks. If it isn't temporal, it can't be causal.

But it turns out also that clock synchronization is one of the fundamental problems in distributed computing. There's a small set of problems that people consider to be fundamental. How do you vote on values when up to a certain fraction of them can be faulty? That's what triple modular redundancy comes out of. How do you get agents to agree on certain actions in the face of some faulty ones? That's the Byzantine generals' problem. Clock synchronization turns out to be another of these fundamental problems. There's a lot of depth in it.

> ► After you got your first position at Cornell, did you continue to work in the same area?

Yes, I generalized the problem to agreeing on all kinds of things. That pulled me into some issues of applied topology, of all things.

I also started looking at the issue of decentralized control. This was a long time ago, when we didn't have the Web yet, but our notion of distributed services had been developed. At that time you had file servers, name servers, printer servers, things like that. I was interested in how you built control software that could be layered on top of a network and programmed at a high level.

I developed a reactive decentralized control system called Meta with a colleague of mine at Cornell named Ken Birman and his postdoc Robert Cooper. We ended up putting it on top of some software that Birman had developed, called Isis, and we commercialized it. We started a consulting company and started producing software for Wall Street. It was also adopted, for example, by some air traffic control systems in Europe. Later we sold the company to Stratus Technologies.

> ► You moved to San Diego in 1993. What was the incentive for doing that?

Family. My wife was working at Rochester, which is a long way from Cornell. We wanted to be in same city and we started looking for jobs. I grew up in Torrance, California, so it was a lot like moving back home.

> ► You were also at the University of Tromsø for a while. How did that come about?

A professor named Dag Johansen pulled me into the problem of fault tolerance in mobile agents. We graduated a student together, who is now working at Microsoft in Oslo.

> ► How did you like Tromsø? Was that your first time in Norway?

It was my first time in Norway. I loved Tromsø. It's very different. The northern Norwegians will tell you they're nothing like southerners. Their accents are different, even to my ears when they're speaking English. And they love their aquavit!

▶   **What kind of research have you been doing at San Diego?**

Well, I've been at San Diego a long time. When I first arrived I was trying to look at Isis-like systems for real-time computing. I produced a system called CORTO. My goal was to separate the real-time systems aspect from the mathematics aspect in scheduling issues. Especially at that time, the world was divided between hard and soft real-time systems. The soft real-time systems do the best they can, which pushes all the deadline problems to the programmer. Hard real-time systems give you a guarantee: If they say they'll meet a deadline, they will. However, they might reject some things that are computable. That puts math problems into the system itself.

I was trying to make a system that would allow you to account for time so that you can do the scheduling externally. It's a kernel of a real-time system that lets you put schedulability on top of it. I thought that was the right approach for distributed systems.

Another thing I'm playing with now is the problem of how to get around the fundamental limitations of replication protocols. People have programmed Byzantine agreements to be tolerant of attacks for some time now, but there is a lack of convincingness to these protocols. You can't have more than one-third failures. If you think about an attacker trying to take over a system, it's hard to see how the attacker could be limited to one-third of the system, because attacks can spread. I'm still working on ways to get around that limitation.

▶   **When did you move to the National Science Foundation? What were your reasons for deciding to do science policy for a while?**

I started in September of 2010. I had been chair of the computer science department at the University of California, San Diego (UCSD) for five years before going to the NSF. Five years is a long time to be chair. While I had kept my research moving forwards, I realized that as a chair my research had gotten very narrow because that was all I had time to do. I wanted to have a chance to re-broaden myself, see science on a larger scale than I could at UCSD. The NSF seemed to be the right place to do it.

▶   **Has it lived up to that goal?**

Oh, yeah, no doubt about it! I didn't realize how much time my job would take. So my research is still moving forwards slowly, but at least in my spare time I'm seeing lots and lots of good science. I'm getting a much better idea of what's going on in the nation and the world in computer science. I feel as if I have a much better grounding in reality, if that makes sense.

▶   **What have you found out about that you wouldn't have otherwise?**

At the NSF I've gotten involved in the National Cybersecurity Strategic Plan. This is something that all the federal agencies with a stake in cybersecurity have gotten together on. It hasn't changed my research, but I have a broader understanding now of what is driving cybersecurity forwards, what the industry is doing, what the government sees as important, what important results are coming out, and where we should be going. As a society, we need to do things that are more reproducible, to make our understanding of cybersecurity less of an arms race and more something we can analyse.

▶   **Let's talk about the US Ignite project. Has it moved forwards since the Oslo meeting?**

Two things are going on. One is that we're still trying to set up the necessary non-profit to be able to announce it. We're making progress but it has been slow because it requires identifying key people who are willing to put time and energy into it. That's always a balance. You can find people who are willing to do it but don't have the pull you want and you can find people with the name but not willing to put time into it. You've got to get the right person and that takes time.

The other thing is that we've been funding some of the initial projects. None have been announced yet, but they will be soon. I know that for a fact. We're still working through details of how we're going to make the public announcement. [Note: As mentioned above, the announcement took place on June 14, 2012.]

▶    Was US Ignite originally your idea?

It was not my idea. The original idea for it came out of the Office of Science and Technology Policy, in the White House.

▶    When you came on at the NSF, was it with the understanding that this was something you were going to be working on?

No, it happened after I joined the NSF. We first started talking about it at the end of 2010.

▶    What made you want to get involved with this?

There are two ways to answer that question. One reason is that the system it's being built on, called GENI, is under my division. I have a responsibility because they're using my resources. The other reason has to do with the very recent growing availability of high-speed symmetric broadband to businesses and residences. Before I came to the NSF, Larry Smarr and Ed Lazowska wrote a document called 'Unleashing Waves of Innovation', which made an argument for the need for fibre to the home and the need for universities to be involved with deployment. In reality, though, the deployment happened without university involvement: The Recovery Act provided the National Telecommunications and Information Administration with $4.7 billion to establish the Broadband Technology Opportunity Program to increase broadband access and adoption; provide broadband training and support to schools, libraries, healthcare providers, and other organizations; improve broadband access to public safety agencies; and stimulate demand for broadband. This is providing the infrastructure and some innovation, but not enough. I see US Ignite as taking the second step that was advocated in 'Unleashing Waves of Innovation'.

▶    Can you tell me how GENI originated and what are the main ideas behind it?

GENI stands for Global Environment for Network Innovations. It's a virtual laboratory that allows you to deploy and experiment with Internet protocols at scale. The easy way to describe it is this: Suppose you want a research internet to experiment with a whole new architecture and not just the current Internet architecture. Many countries have research internets, including China and Japan. In the US, however, we have built it as part of the regular Internet. What enables this is something called slicing, a way to carve out part of the routers and bandwidth so that you can do experiments in a way that's isolated from the rest of the network.

It uses, among other things, a network mechanism developed at Stanford called Open-Flow. With a typical router, data forwarding and data routing --- what are called the data path and the control path, respectively --- are performed by the router. With an OpenFlow router, the data forwarding is done as normal, but data routing is done by a separate programmable server. This allows one to, essentially, deploy new switching protocols that can run in parallel, using the same switches. GENI builds on this to provide an abstraction of a network slice, which is a set of virtualized network resources that can be reserved and programmed by an experimenter.

▶  Who is using this experimental Internet?

It's mainly individual university researchers using GENI. That's what the NSF supports. We have a GENI Engineering Conference, GEC, which brings together people who are building GENI as well as using GENI. The 13th meeting is happening in UCLA next month, and I will be there. The demos are always interesting to see.

▶  Can you give an example of a particularly creative project that uses GENI?

Stanford University has built a great demo that shows how they can use GENI to do seamless handoffs between their campus WiFi and WiMAX networks: They drove a golf cart around the outside perimeter of Gates Hall and maintained excellent connectivity as their mobile device used different base stations. Stanford also has a good demo of their Aster*x software, which uses global server load and network congestion to dynamically route and load-balance the network paths being used by a flash crowd of service clients. TransCloud is another GENI project: It is a highly geographically distributed environment that supports services on highly distributed processes. It, too, makes used of GENI's dynamic networking for high performance.

▶  How does US Ignite differ from GENI?

From an implementation point of view, US Ignite is enabled by GENI. GENI brings the ability to do sliceable networking to universities. US Ignite will enable cities with fibre to the home to participate in these at-scale experiments with the applications developed by university researchers or by individuals supported by local organizations.

▶  What can cities do that universities can't?

There are two things cities can do. First, you can have real people using your services. It's not that students aren't really people, but it's a different environment that offers different opportunities. It also allows a much broader base of developers, including small companies, entrepreneurs, and high school kids who are bored at home but have great ideas. US Ignite is a way to allow broader experimentation. It also encourages public–private partnerships, in health care, emergency response, public safety, education, and so on.

▶  To what extent do you have ideas already about the applications for fibre
     to the home and to what extent are you just putting the technology out
     there and seeing what will come of it?

It's neither of those two, but closer to the second. We've been holding a series of workshops to get people thinking about what you can do with this environment. The NSF, in some sense, is a bottom-up organization. We're not in the business to tell people what to build but, rather, we are supposed to provide an environment to let innovation occur.

But you've put your finger on an issue, because this project is such an unusual thing. Having so much bandwidth available, what are you going to do with it? It's not quite a 'build it and they will come' situation, but we will help people talk about what to do with it.

▶  How many workshops have there been so far?

Three. They will continue but we're at the point where we're funding some of the proposals that came in as a result of these workshops.

▶  Can you spill the beans about any of them?

I can't tell you about specific proposals until they clear our funding. I can tell you what people have been talking about and I talked about them in Oslo. Emergency response is one that people seem quite interested in, because of the disaster in Japan, the earthquake in Haiti, and the memory of Hurricane Katrina in New Orleans. The thing people seem to be focusing on here is the combination of collecting information, crowdsourcing, and having access to backend computation so you can do modelling. It's not so much sending out alerts saying a hurricane is coming, but collecting information at different stages, pulling that into a machine that can compute, and then propagating that information out to the public.

Other examples that have come up are in the field of healthcare. Obviously, the more bandwidth you can have, the more interaction you can have with doctors at a distance. It's impressive what can be done with the patient at one end of the fibre, the doctor at the other end, and a nurse who can mediate. It's like a remote clinic.

▶    Are some of these in existence already?

Georgia Tech has done some work on remote clinics, which I saw last year at a workshop called Broadband 2020. And Case Western Reserve University is doing some of this now in Cleveland. They have applied it to the care of people with type 2 diabetes, a problem all across the United States. Some people are uncomfortable about going all the way to the Cleveland Clinic. By putting in remote satellite clinics, connected with high-bandwidth fibre, you let these patients bring in their families and have the security of being in their own neighbourhood, yet at the same time have world-class physicians working with them.

▶    Can you talk about the idea of public challenges and whether the NSF will be able to work successfully with this model?

I think so. The whole method of challenges is new to the NSF. It's only since I've joined that we've gotten permission to do these challenges. We're not just going to do it for US Ignite, by the way; we will also do it for cybersecurity.

Ever since the Red Balloon Challenge by DARPA that generated so much excitement, it's become clear to many people throughout the government that these kinds of challenges really excite the public and get people involved who aren't part of the normal research infrastructure. Also, they help make the public aware of what we're trying to do. It builds excitement as well as comes up with innovative solutions.

▶    Do you have any sense of how much interest there is in the Mozilla Apps Challenge?

It hasn't been officially announced yet. Remember, we're trying to get this whole thing announced still, which is taking time.

▶    You've mentioned the gigabit project in Chattanooga and Google is doing something similar in Kansas City. Is this best done by the government or by private enterprise?

And the answer is… a partnership! A public–private partnership. When we announce US Ignite, you'll see that there is a lot of industrial partnership.

▶    Is it realistic to expect a huge number of people, from high school students to local entrepreneurs, to come up with applications that work and that are new and different?

There's an awful lot of creativity that goes on with young people at universities! If you ever go to any of these hack weeks – Yahoo has them and Qualcomm has done them – they're basically events at which students spend a week or a couple days to develop new innovations. It's always amazing to see what they come up with.

> Looking 10 years into the future, what will the big challenge be? It seems to me as if US Ignite is more of an intermediate-term challenge, over the next five years. Suppose US Ignite is successful? What will be the challenges after that?

That's a really good question. All I can do is hazard a guess, I must say. It seems to me that one of the great things that high bandwidth enables is person-to-person communication at a very high fidelity. I image that US Ignite and fibre to the home have the possibility to change how we work and interact with people. I don't have the wisdom to say exactly what it's going to be like.

It's certainly the case that we spend a lot of our time and energy commuting. It adds to pollution and global warming. Once we have very good points of presence or protocols that let you work at a difference at very high fidelity, then our need to get together and meet is going to change. Eventually, I think that will be a good thing, although right now it sounds a little bit scary. Imagine what a world would be like where everyone telecommuted. Would everyone become detached because they are no longer meeting around the water cooler? I don't think so. With very high bandwidth, it may just become a virtual water cooler. Probably it will be a hybrid thing where you work at home a lot but also go to meetings when you need personal interactions that you can't get at a distance.

It's going to change the way our cities work, our commuting patterns. I think it will give power back to communities that are pleasant but where people won't live because it's economically unfeasible. For example, I have a friend who runs a tour business in Japan and he lives in a remote part of Japan. Yet this is a society where everyone is gravitating to the major cities. Cloud computing makes it possible for him to run his business from a distance. Fibre to the home will make a lot of things feasible. It's pretty exciting!

> Just out of curiosity, what did you think about Simula on your visit? Can what they're doing be replicated in the US?

I had never been to Simula before, though I had certainly heard of them. I was very impressed with the depth of what they're doing. I'm hoping to learn more about them, because it's an intriguing model. We have centres at the NSF, the Software Research Corporation, and Industrial/University Cooperative Research Centers (I/UCRCs). But it seems that Simula is one click up from that in terms of size and control. When I look at I/UCRCs, the people there are primarily faculty of their universities. When you're at Simula, you're at Simula. It's more like a national lab, like INRIA, but they're pretty nimble and it isn't clear to me how they are doing it.

> How long do you plan to stay at the NSF?

I'll be at the NSF for three years. I do intend to go back to UCSD. What I'll do there I don't know. I'm just now thinking about it. When I left I was a chairman, doing individual research. Now that I have spent time at the NSF, it will give me an opportunity to perhaps define our own large-scale research groups at UCSD. We can perhaps do more for undergraduate education. I knew that my department is strong, but I have learned some weaknesses, some things that we can do better.

> Do you have some ideas that are applicable to other universities as well?

We all need to break out of our stovepipes. It's easy to be an expert in one area, but in computer science we're an enabling technology. There's a lot that we can offer by breaking out of a narrow point of view and crossing multiple disciplines, whether they are economics, sociology, medicine, or linguistics. I like to say that in computer science we value the scientist who sees low-hanging fruit where other people don't even see trees. Universities that understand the value of breaking out of the traditional research boundaries and learning how to build on each other will be more successful in the long run.

There are changes we need in education as well. We could talk about these issues for hours. One of the many issues has to do with teaching cybersecurity. For example, there is a programme started by the National Institute of Standards and Technology called the National Initiative for Cybersecurity Education, which aims to broadly teach cybersecurity principles, starting in elementary school. You can compare it to a question of public hygiene. We teach kids to wash their hands, brush their teeth, and cover their mouths when they cough. There is some actual science behind these practices, but they are simple things we teach our kids to help them lead a better life.

We can also conceive of a cyber-hygiene that needs to be taught. There are some things we need to teach at the lower divisions, as in K–12 education, but it should also continue at the college level. For example, what risk do you take on when you borrow someone's code? What does it mean to have a system that is robust? How can you write code so that someone else can pick it up after you? When I learned computer science, you wrote things from scratch. Now you build software on top of layers and layers of software, all provided by different individuals. If we are going to produce engineers who know how to design systems that are more robust and more resistant to cyber attack, we need to do more than teach security as an upper division course.

# The Internet of Things

### An Interview with Heinrich Stüttgen by Dana Mackenzie

Heinrich Stüttgen, the vice president of NEC Laboratories Europe, brought a unique and valuable perspective to the Challenges in Computing conference, as the only speaker from the commercial sector. His topic, the Internet of Things, was, however, quite familiar to the other attendees.

The idea behind the Internet of Things is that each of us owns more and more devices that contain possibly multiple sensors, which can be used not only to inform us about the rest of the world, but also to inform the rest of the world about us. Smartphones contain GPS locators and accelerometers; automobiles contain sensors that can interact, for example, with emergency services or with toll booths; electric 'smart meters' can track our power usage and inform us (and the utility company) about the cost of our energy usage in real time; and many industries use radio-frequency identification (RFID) sensors to track shipments or inventories.

What all of these current-day applications have in common is that they are all islands, 'intranets of things', rather than one connected 'Internet of Things', as Stüttgen put it in his lecture. However, the day is coming soon when these applications will begin to go online. The size of the Internet will suddenly grow from 1 billion people (estimated) today to 2 billion people and 5 billion phones and 2 billion RFID tags and 1 billion cars. And that is even before you start counting the smart teapots and the smart toothbrushes. 'According to Wikipedia, 50 to 100 trillion objects could have sensors', Stüttgen said.

What will all this information be used for? Can it improve our everyday lives? How do we keep it from impacting our lives in a negative way, for example, by loss of privacy? These are some of the challenges for the future that Stüttgen identified. On a more technical level, the principal challenges are complexity, heterogeneity, and integration. 'We need to have a common infrastructure that allows for communication between heterogeneous devices', he said, 'And we need interoperability at the application level, not the device level'. NEC has developed a model for such an infrastructure, called Isis, which Stüttgen mentioned in his talk and discussed in our interview.

The interview took place on January 31, 2012.

▶  First, tell me a little about your background. What was your original expertise and what do you do now at NEC?

How much do you want – five seconds or five hours? Briefly, I studied computer science at Dortmund, Germany, and then spent a year in Buffalo in the US, where I got my master's. I went back to Dortmund to get my PhD. All of the time I was working in operating systems research, specifically memory management. I got a PhD there in 1985 and then started to work at IBM. I spent two years at their development lab in Böblingen, near Stuttgart, then I went to their European Networking Center in Heidelberg, a small network research centre. I worked there for 12 years and eventually became a group manager.

A. Bruaset, A. Tveito (Eds.), *Conversations about Challenges in Computing*,
DOI 10.1007/978-3-319-00209-5_3, © Springer International Publishing Switzerland 2013

Then a headhunter found me and I moved to NEC. I started a new research lab, which was also in Heidelberg. I remember that on the first day I was sitting on the floor because there was no furniture yet! It came in the afternoon of the first day.

The laboratory has grown quite a bit. We're up to 100 people now. I tell people that our focus is on networking, not communications. When people hear *communications*, they think of wireless and optical. We're working on the software layers above the physical transmission, from communication protocols to the services that run on top of these.

> ► How much is what your lab does dictated by the company and how much are you free to pursue your own interests?

One of the things I've enjoyed at NEC is that there is a lot more delegation of responsibility as compared to my previous employer. We are able to design what we want to do. Of course, our projects have to fit the company's interest and there's coordination between the various labs of NEC. The yardstick is whether we can develop new technologies that have good potential for contributing to NEC's future business.

It's a little bit different from an academic lab, where the main purpose of your being is to write publications, sometimes almost independently of whether anyone reads them or not! In an industry lab it's all about impact. Not all of that is always seen by the public. You might learn much later that there is a new product, but you have to look carefully to see what the researchers contributed.

> ► Can you give any examples of products of NEC that came out of your research lab?

A typical example is in security technologies. We've been working on an advanced form of identity management. We recently transferred this identity management system to two different business units within NEC Japan, where it was integrated into existing products. In both cases you would think that this product has been around for a while and you wouldn't necessarily realize that some of the technology included came from our unit.

We have also been working with our display company that sells the professional displays I talked about when we met at Oslo. We're trying to do something a little more advanced than just a plain display and we eventually decided to integrate some of our machine-to-machine (M2M) technology into the display product. So now the display can react to the person standing in front of it. In that case it's fairly obvious that the sensor technology that we've integrated into the display is our piece. Sometimes you can tell and sometimes you can't.

> ► What are the customers using the displays for?

The product is not in the market yet, so we'll still have to see. The displays can be used to target your advertising to the person standing in front of it. We had a demo where the display would identify you from your PDA or your smartphone using a Bluetooth connection. It would basically be able to read your shopping list if you allow it, direct you to a place within the shop to buy your goods, and print a coupon for you that you can take a photo of with your smartphone. You can do all sorts of things by interacting with those displays. That, of course, is more fun than looking at some prescheduled content and at the same time there is a lot of money to be made through better-targeted advertising.

> ► It reminds me of the movie Minority Report, where the main character was seeing ads based on who he was. But in the movie it didn't seem to be a good thing.

I don't remember the movie, but there are privacy issues here. I've seen cases where companies install a video facility in order to do age and gender recognition of the customer, a very useful approach for targeted advertising. However, when you install video facilities in

Germany, you have to be very careful to respect privacy laws. This issue is very different from country to country, even within the EU.

> ► You mentioned in your talk that it took you a very long time even to install a camera in your lab.

It's very specific to Germany. In a public place, if you install a camera, you have to have government permission. You have to warn people in the area that they may be observed by a camera. This may be an alien thought when you go to the UK or the US, because I think the privacy laws are way less strict there than in Germany. There are some technologies you may be able to use in all countries and others where you need to be very careful regarding privacy aspects on a country-to-country basis. Some of these concerns are well founded and others may be a bit exaggerated. Frankly, I don't understand why I need to have government permission to install a camera that is active only when the lab is closed, when nobody is supposed to be there.

> ► What was the reason you wanted to install the camera?

We had a security incident and we never found out what actually happened. This led to the thought that we should put in cameras or motion detectors to see what happens when the lab is closed. After a one-year administrative process we were allowed to do it legally.

> ► Will it seriously affect the market to have so many different regulations on privacy? Will NEC eventually negotiate with the governments?

That's a difficult question. Sometimes the problems go away as a technology develops. Sometimes the technology develops faster than the legal framework. Take copyright, for example. Eventually people may realize we want to preserve privacy but we don't want to ban the use of technology for good purposes. Whether NEC would take a role, I don't know, but I doubt it, at least not within Europe. From my personal view, Europe needs to harmonize its privacy laws urgently. As Europe more and more becomes a single big country with multiple nationalities, a single marketplace, you need to have a comparable legal framework everywhere you do business.

> ► Let's move on to the Internet of Things. You said that if you ask five different people, they will give you six definitions of what it means. Do you have a favourite definition or maybe two?

Basically, the Internet of Things says there are things, sensors in technical terms. Internetting means that they are talking to each other. An infrastructure that allows that to happen is what I would call an Internet of Things. Some standardization bodies, such as the European Telecommunications Standards Institute, talk about M2M. Not so long ago, people used another term, *pervasive computing*. Another neat term is *cyberphysical systems*, because the cyberworld interacts with the physical world.

> ► What's the value of the Internet of Things to consumers? It seems as if you were talking more about the value to businesses in your talk.

We are actively researching a combination of robotics and the Internet of Things. You could have all sorts of helpers for assisted living. One demonstration we've recently done, originating from a European-funded project called Florence, was about using this technology to help an elderly person in her daily life. The robot would monitor whether all the windows are closed or whether one is open. The robot doesn't have to close it; it just has to tell her to get up and close it if you don't want to get cold. If a nurse comes in, you can identify her with an RFID card, another type of sensor. All of these systems need to communicate with each other.

Assisted living is a big area where this can be used. If you look at the demographics in the industrialized world, whether it's Europe or the US or Japan, people are getting older and older. We will definitely need more help to enable these old people to live independently.

►    **Have you tried this assisted living application in practice and how have the people using it responded?**

No, it's still a little bit early. We collected feedback on the demonstration. Care organizations, such as the Red Cross, would be very interested. If I discuss this with my partner, who is a psychotherapist, she might have an entirely different view. She would tell you that what old people need is human interaction, not interaction with a machine. Our goal is not to replace the nurse, but help the nurse to do her job efficiently and well.

►    **Does NEC have other projects related to the Internet of Things?**

There are different streams of projects. First, there's a certain set of projects trying to develop the basic platform for doing all these applications. Sensei was one of our earlier projects in this area, where we basically built an intranet of things. It was a closed network, combining different sensors for specific applications. [Note: The Sensei project had about 30 partners, including universities and other telecommunications companies.]

Now we're trying to do a new thing, turning the multiple intranets into an internet. That takes more platform work and a big challenge is managing the complexity of a multitude of different devices. Every day there is probably a new type of sensor around and you want to have a software structure that incorporates it into the Internet of Things very efficiently. You don't want a program that takes you five weeks to integrate one new type. We have internally an implementation where it took about two minutes to integrate a new type of sensor.

A specific issue that leads a different stream of projects we've been working on is security and privacy. When you think about security, typically you encrypt data. Mathematically, this is a very complicated process. If you have a small sensor, you probably can't do 128-bit encryption on it; you would not be able to do it in time. Can we introduce a reasonable amount of security and privacy without exceeding the capabilities of these devices? This 'lightweight security' doesn't give you perfect protection but it closes the door to the point where not everybody can screw up the system easily.

The next set of projects consists of specific applications. I already mentioned the Florence project, combining M2M technology with robotics for assisted living scenarios.

A totally different scenario is a new project called Campus 21. In this project we are looking at how to use sensor and M2M technology for building energy control. M2M technology can help you to drastically save energy, but you have to consider the following: Do you have to deploy those applications in-house or can the cloud provide these services in the background? Do you want your neighbour to be able to see your energy consumption or do you want some privacy here? You have a lot of these issues that you have to look at specifically for each application.

►    **Who are the people who started the idea of the Internet of Things?**

One of the pioneers is a professor from ETH Zürich, Friedemann Mattern. Mattern gave many early and very inspiring talks about pervasive computing and its possibilities. I could mention some other names, but it isn't quite fair because a lot of people did a lot of experiments. The University of California at Berkeley developed some sensor devices, which they called motes. These were quite important because they were not expensive. Also, you could put software on them. In one of our earlier security projects, called Ubisec&Sense, we bought some motes and implemented lightweight security algorithms on top of them. The availability of these programmable motes has kicked off a lot of interesting experimental work towards the Internet of Things.

Although RFID chips have been around for a while, they're not programmable. People like to talk about what can we do with RFID, but it's only an identification technique; it's not a processing technique. So RFIDs didn't quite inspire researchers quite as much as more intelligent sensors, at least not outside the logistics area.

> You mentioned something called Isis, which you described as an infrastructure that allows the intelligent processing of data. Could you tell me more about it?

First of all, *Isis* is not an acronym, just a name. It basically is a software architecture and software platform to integrate different types of sensors and provide some basic processing infrastructure. It doesn't do applications but it provides you a logical interface to different kinds of sensors, so that the programmer doesn't have to worry about the physical infrastructure. The application programmer would request certain sensor data, Isis would break it down and distribute the request to different sensor types, collect the information, and provide the result.

Isis was motivated by the fact that the key inhibiting factor is software complexity. Our idea was to have a common platform where, from the bottom, we can integrate new sensors but the application programming interface is basically stable. The application programmers will just be happy to get new data provided by new types of devices, but they won't have to communicate with them directly.

> How far along is this Isis platform?

I think, basically, it's ready at a certain level. We are actually using it not only in prototypes but also in the display project I mentioned earlier. So far we've been using Isis only with a small number of sensors, meaning anything below 1000. We haven't really tried any experiments with millions or billions of sensors controlled by Isis. With that many sensors, the resolution and discovery of devices are yet to be resolved.

> Will it be possible for other companies to buy into it and use the Isis platform for their projects?

Yes, of course, that's a commercial decision. As a research lab, we collaborate with other companies as part of the European framework. For instance, we have used the platform in the retail lab of SAP. Can they buy it? Yes, we're open for negotiation, but there's not a product with a price tag that you can order from a menu at this point in time.

> In your talk, you identified four main challenges for the Internet of Things: complexity, heterogeneity, integration, and privacy. We've already talked about the privacy and integration. In any of these four areas are there other aspects you'd like to mention?

The issues of complexity, heterogeneity, and integration are closely related. Because things are heterogeneous, it becomes complex to integrate them. The other aspect is the sheer volume. If you have zillions of devices, producing data every minute, you have to deal with huge streams of data. There are different dimensions of complexity. We may not have all these dimensions at the same time.

To take an example, if you wanted to have up-to-date information on actual traffic movement in New York City, you could have sensors in every car. That would be quite a lot of data to deal with. For another example, your refrigerator at home might have only a few sensors in there and the complexity may not be very high at this point in time.

> What kind of sensors would a refrigerator have?

At a minimum, temperature and a door open/door closed sensor. You could have devices that read data from RFIDs which would tell you that your milk will expire in two days. You could get arbitrarily fancy in that regard. Ideally, you would have a little printer that prints out a shopping list and sends it to your supermarket for you to pick up when you come home.

► What talks at the Challenges in Computing conference struck you as important?

It opened my mind to look at some areas I haven't dealt with in the recent past. Omar Ghattas talked about the seismic project in Austin. That was a totally different area than what I'm working on, but there is some relationship. If you do seismic processing, you're probably dealing with sensors again. It showed some applications we hadn't been thinking of.

Also, Natalia Trayanova gave a talk about the heart modelling, which was totally different and quite inspiring. I was wondering is it realistic to have this on a broad scale for everybody? Or is it going to be too expensive? But she may have had exactly the same thoughts about what we're doing. We're doing research here, trying to create possibilities. It's too early to ask whether certain things are going to be realistic in terms of price. First of all, we have to understand and hopefully solve the technical side of the problem. I personally find it rather dangerous when people try to shoot down a new idea just because it won't fly commercially from day one.

► Does that happen more often in the commercial world than in academia?

That's a very difficult question. I think what you see more often – not talking about NEC specifically, but in general in industry – is that people understand their current market well and if a new idea doesn't fit into their picture of the market's evolution, they become reluctant to invest in it. Not everyone is gifted enough to take a different perspective and see new longer-term opportunities.

► What about in academia? Do academics also have their own little boxes?

Oh, boy, that's another difficult question! For some academics, but certainly not for all, if they can write an interesting paper about a technology, that is enough justification to do the research. That's probably not enough for an industry research lab. Research managers want to see technologies contribute to the future business of the company. Based on that different attitude, if you're just interested in whether you can write a new paper, maybe you're more open to new ideas. If you're thinking about whether you can make a new business with this, then a lot more constraints are in the way. I think it's good and bad at the same time. It's good to consider the utility of technologies, but if you restrict your ideas too early in the process to the ones you can turn into money, then many good ideas will never fly.

In that sense it is probably important to have university and industry research with different boundary conditions, allowing one to explore the future in different directions and thereby not lose too many good ideas on the way.

*Heinrich Stüttgen gratefully acknowledges the support of the European Commission within its Sixth and Seventh R&D Framework Programmes, which has supported NEC's collaborative research activities in the projects described above (Florence, Sensei, Ubisec&Sense) and many more.*

# Part II
# Computational Science

# The Mathematics of the Mind

*An Interview with Hans Petter Langtangen*
*by Kathrine Aspaas*

*'Newton said: If I have seen further it is by standing on the shoulders of giants. Today we are standing on each other's toes'*
Richard Hamming (1915–1998), American mathematician

What is a professor? The day's first really good question comes from Hans Petter Langtangen. Naturally, he directs the query to himself and is obliging enough to supply the answer – and in no short measure.

"A professor," he says, "holds the highest possible level of formal scientific competence. It means that you have expert knowledge in a field of study, with a breadth and depth which, in the natural sciences and medicine, normally corresponds to three doctorates. You must also, over a 10-year period, show that you can be a driving force to discover something new and important that no one has known before, and get this new insight published in internationally recognised scientific fora. Such competence is a good starting point both for teaching at a university and for training the scientific researchers of tomorrow. It concerns me greatly that the role of the professor is changing radically. Previously a professor was measured in terms of independence. Today we are measured in terms of collaboration. To use a football metaphor, you could say that universities used to hire mostly centre-forwards, which results in a lot of prima donnas – and the same problems they have at Real Madrid. The club may have the best players, but they don't always have the best football team. What we need are players with different qualities, so apart from me there cannot be too many prima donnas if we are going to be productive."

With that the tone of the conversation is set. There is no lack of humour or self-awareness in this man, who – by the way – grew up a mere stone's throw from here, from the place where he studied, where he teaches his students, and where he is now sipping the day's first cup of coffee: the University of Oslo. The plan was to take over his parents' city-centre jewellery and watch shop. But it was his sister who ended up as the watchmaker. Langtangen himself was ensnared by mathematics and physics, and in addition to words like *omnifarious*, *nerd* and *musical*, it is his ability to see the big picture that is often highlighted when reference is made to this shopkeeper's son. Quite simply, he has good intuition.

"A great deal of what I observed when I studied here as a 20-year-old has been crucial to my way of working," he explains. "Something as simple as the importance of the computer, which I was introduced to back then in the early 1980s. Like many others I realised that it would revolutionise mathematics and physics, but I also realised immediately that it would revolutionise how these subjects were taught. Unfortunately it would take 30 years before that latter realisation became an academic truth. These are extremely slow mechanisms, because many people don't want change – particularly not when they're teaching."

▶ The mathematician Richard Hamming has said that it was widely believed in the 1930s that mathematicians didn't need calculators. They were so clever themselves.

A. Bruaset, A. Tveito (Eds.), *Conversations about Challenges in Computing*,
DOI 10.1007/978-3-319-00209-5_4, © Springer International Publishing Switzerland 2013

"Too true! It's been like that the whole time. No mathematician worth his salt would use a computer. Even today it's pencil and paper-based mathematics that gives honour and status. If I were to tell you how we teach…"

▶    When you say it like that, I'm not sure I want to know.

"No, I don't think you do. And you would certainly not be allowed to write articles on it. There'd be riots in the streets."

▶    But what is the reason for all that?

"Ah, well, that's a big question. I've been pondering this for many years, and it's not unique to this place. It's the same all over the world. Why is the teaching often so out of date? We are keen for computers to develop curricula and text books, but they've almost never done so anywhere – apart from in the computer science field itself (THIS SENTENCE IS STRANGE?). It's very strange, because everything in mathematics is extremely rational and logical. But when it comes to reforming the teaching of it, it seems as if the process is governed by laziness and conservatism – not rationality."

▶    Could it also be down to fear?

"Yes, because many teachers do not master the new computer-based methods. It's paradoxical that academic circles should be so hugely conservative at the same time as everyone is recruited because they are highly creative in their research. It just doesn't hang together."

▶    The philosopher Arthur Schopenhauer said that all truths pass through three phases: first they are ridiculed, then they are violently opposed. And finally they are accepted as being self-evident. Do you think that is inevitable, or is there something that can be done to change this culture?

"I am part of a group of professors here at the UiO who have actually managed to reform how we teach, largely by collaborating in a way that no one has succeeded in doing before. The generation of professors before us were fiercely competitive one-man shows. When I was a student we moved effortlessly between the mathematics building on one side of the road and the physics building on the other, but our professors very seldom did. The gulf between was far too wide, and they spoke only disdainfully of colleagues with offices across the street, even though they were working on practically the same things. But when I turned 40 I made new professor friends, and they are each so clever that they don't need to compete. It's a matter of confidence. We back each other up, and can therefore get things done across territorial boundaries."

▶    Have you and your colleagues started using computers, then?

"You might well laugh, but it has actually made us world leaders. It's 30 years since the computer became a household commodity, and I'm starting to get optimistic about teaching reforms. All the textbooks are going to be rewritten. What the textbooks say today is based on what we can do with pencil and paper, and they are very primitive tools, even though using them well requires a very advanced intellectual level. We can no longer pretend that computers don't exist. Their superiority as number crunchers is far too great, even though a small minority of people are very good at working things out on paper."

At this point we leave teaching methods and move on to the actual research.

# Everything is Mathematics

Hans Petter Langtangen has been associated with Simula since 2001. The objective was to create a flexible organisation engaged in outstanding research with obvious significance and applications. Today he heads Simula's *Center for Biomedical Computing*, a prestigious centre for outstanding research that is home to a melting pot of mathematicians and computer people, physicists and clinicians. There are students, postdoctoral fellows, associates and permanently employed researchers. Together, a dynamic group of 20–40 people. And at the end of the day there is one fundamental question that drives him: how does the world work?

"We possess a huge amount of knowledge and understand a great many exciting things," he says. "But it's really only about this one question: how is the world put together? Nature? I try to understand it through equations. It excited me the first time I understood it was possible. It still excites me. Just a simple thing like a football that is kicked. The flows in or around any moving object. After 30 years I'm still head over heels in love with throwing a ball and describing it mathematically. The pattern of knobbles on a golf ball and the reduced resistance it produces."

▶    So it is this fundamental knowledge you use in your research into biomedical issues?

"Yes, we want to help doctors make diagnoses and develop treatment methods by looking at what happens in the body through the prism of physics. By following electrical signals in the heart and the flow of blood in the brain we gain a fundamental insight into how the body works – or doesn't work. Take strokes, for example. Around five per cent of people our age have aneurisms (small dilations of the blood vessels). Like overfilled balloons, these dilations can burst, causing the most serious kind of stroke. We want to find out if there is something in the way the blood flows that leads some people to develop aneurisms. It could be due to the blood vessels' geometric shape, resulting in incorrect blood flow. In this way we can help doctors decide whether or not to perform surgery. They don't want to go into the brain unless it's absolutely necessary. After many years of research we've discovered that high speed and strong vortices in the blood can be dangerous, and that this is linked to the geometry of the blood vessels. When the blood starts swirling about and behaving chaotically, turbulent currents can arise. The objective is a nice, laminar flow, and we now have extremely accurate calculations showing that dangerous aneurisms are often combined with turbulence or very complicated flow patterns. This is still a hypothesis, but it looks very promising. If the hypothesis proves correct, we could envisage a diagnostic method that seeks to reveal turbulent blood flows in the brain. If you have a turbulent blood flow combined with aneurisms, for example, well – then you elect to perform surgery."

▶    Where does neurobiology and neurophysics come into this picture?

"We're just looking at the physics, but these things are closely connected. An aneurism occurs when the wall of the blood vessel has weakened, and the cells in the blood vessel wall are extremely sensitive to blood flow. High blood pressure, for example, causes the cell wall to be reinforced. There is a lot of chemistry and molecular biology involved here, also connected to electrical signals. The big, open question in the natural sciences today is the linkage between scales. The major challenge is that we have processes that occur on many different scales. We have electrons and atomic nuclei at the one end, and galaxies at the other. Nano-science at the one end, and astrophysics at the other. When we set out to study a phenomenon, for example a football, we can collapse it to one single mathematical point that moves. I can ignore completely the fact that there are electrons and molecules in the football, and this simplification is amazingly powerful. The natural sciences are largely about finding the right simplifications so that we can pick the world apart and study each element separately. We've been doing this for a long time, and we know a lot about the various bits. The crunch question now is what happens when we join the bits together on completely different scales between nano and astro."

► How can you study the brain without the biological aspect?

"We work with the human brain through the prism of physics. We look at it as a physics problem, not a biology problem. But it is important to find the link, and we need biologists and physicists working together on this. That is the challenge."

► It sounds as though we are at a crossroads in science – from studying separate elements to finding the links between them?

"The big unsolved problems of today are about how the individual components affect each other. But if we are to be productive and get results, we must continue to look at the separate elements at separate scales. We simply don't have the tools we need to link these scales together."

► What kind of tools do we need, and how can we acquire them?

"We're waiting for a smart approach, but the problems so far seem extremely challenging mathematically, physically and computationally. In the meantime we must combine the tools we have in one way or another."

► That made me think of the Kavli Award, which is presented to researchers on three different scales: the enormous – astrophysics; the tiny – nanoscience; and the internal – neuroscience. Linking these three together sounds like Nobel Prize-winning material…

"Absolutely. And to have a hope of finally working it out, we have to share. That's partly the reason why Simula operates with open source code in a lot of its research. It speeds up the research by making it immediately accessible for others. Furthermore, we make the research reproducible. In a chemistry or physics lab there are very strict rules for documenting what you do. In this way others can repeat the experiment, and when it is repeated enough times it turns into knowledge. But up until now we've been working in a very primitive way with respect to computer simulations. We don't even remember ourselves exactly how we did the simulations, which makes them impossible to verify. That's why Simula is fanatical about reproducibility. Everything is open, and we provide enough information so that other people can repeat the same work – in other words, verify it – or build on it further."

► What could be the arguments against such openness?

"Usually, commercial considerations. When something is freely available, you can't make money out of it. But we think differently. Our primary success criterion is influence, and we gain far greater influence by having open software and simulations. Even though they are free of charge it requires specialist expertise to adapt the software and the mathematics to a specific problem that the industry is keen to resolve. We have that competence, and when it's used to create a specific adaptation, the adaptation can have a high market value. So our model is to have a lot of open source code, where the volume costs nothing, but charge for developing these magic lines of code that sparkle on top of the basic foundation. Imagine a large box of tools. The trick is to pick the right ones and combine them correctly, so that you can solve your particular problem. That's why I call a carpenter, even though I have a well-stocked toolbox at home.

On top of that there's the pleasure of achieving something of major importance. Take the development of Diffpack, for example. This is a development environment that helps scientists spend less time programming computers, and therefore more time on their core research. When I worked with Diffpack I spent an enormous amount of time programming. While other scientists wrote papers and boosted their H-indexes, often by using Diffpack, I sat and wrote source code. Believe me, it seemed exceedingly stupid at the time. But it has since proved a wise move, because it has had wide-reaching consequences. Other scientists received help to solve equations in a simpler way, and I have received gratitude and recognition, for example by becoming an editor of leading scientific journals. There's a direct logical connection here."

▶     So gratitude is the new unit of currency?

"There is one coinage that dominates the world of research and academic endeavour, and that's recognition. We have it in limitless measure, and that is what is so nice. Because we can use it strategically, at the same time as it is also jolly pleasant. As a young university staffer I experienced that colleagues never said anything if someone had done a really good job. If you didn't hear anything you were doing a good job. If you got a bit of criticism your work was fairly good, and if you received a lot of criticism it was barely up to scratch. The culture today is a lot more generous. When we meet up now we remember what we're supposed to do; that I'm supposed to mention something I thought was really good. That makes the other person happy and they work more and better afterwards."

▶     In some ways that sounds like the physics of recognition?

"If you like. And it works very much better than the so-called H-index, which reduces a scientist's career and achievements to a single number! Of course, other formulae would give other numbers. In my view this is a bad simplification, a long way from the natural sciences' good simplifications. For me it's important to uphold the tradition of 'peer review', where an expert sits down and assesses various aspects of a scientific career or research result. This makes sure that considerations that more adequately embrace the complexity of being a scientist and colleague are taken into account. At Simula we look for those qualities that create great scientists: interest, engagement, unquenchable curiosity and wonderment. Those who wander round with a permanent question in their heads, which they absolutely MUST find the answer to. Preferably lots of questions. On many scales. Who are burning to achieve something. Then we can dream about outstanding research, and Simula has now been through five international assessments – all with top marks – so there must be something in it. To do that we must recruit people who can maintain this level and who possess something extraordinary. The extraordinary lies in this drive. When we find such people they always change the way we operate. The challenge is to find them."

▶     What would you say lies at the heart of your own research? What do you
        want to contribute to the world?

"My research's major contribution has been to make it easier to create a laboratory in the computer, where we can experiment with the world. The number of people who can create this kind of virtual laboratory has increased, and they can do it faster. Predict the future. Estimate development trajectories."

▶     Which brings us back to the connection between different scales. Do you
        think it will ever be possible to describe feelings mathematically?

"Yes, and I don't think it will be long before we can. The Norwegian professor Gaute Einevoll gives a lecture called 'Can we calculate how we think?', and he has an optimistic answer, even though he hasn't found it yet. It's probable that the exponential increase in knowledge stemming from brain research will quickly bring us closer to quantitative measures in areas such as psychiatry. The major structural change in all areas of study comes from our steadily increasing ability to measure all kinds of things. All subjects will become quantitative. It's well underway in medicine, and we can already talk about emotional chemistry. We know, for example, that mating adaptations in various animals, including humans, are based on smell. Species strive to mate with a partner who complements their own genetic material. Not very romantic, perhaps, that so much in this area is purely down to chemistry."

At this point the professor takes the opportunity to slip in a bit of trivia about his own brain – that it needs music, for example. He plays guitar in the Simula band, which also includes a vocalist, keyboard player, drummer and bass player, as well as an entire horn section. They play rock and jazz – often improvised. His own research follows a similar pattern. There is structure and harmony, but he likes to turn things around and put them back together again in his own way. His students call him the *Mystic Mathematician* – an improvising, intuitive lecturer, who does not have time for long, complicated calculations, but prefers to construct the answer the way he feels it to be. One half here and one half there. That is how he constructs the answer through a combination of gut feeling and experience. Can lecture without knowing all the details. Can make great leaps and land on the right answer. But what is really the most important thing he teaches his students?

"I am very practical in my approach," he explains. "I only teach my students things that they really have a use for, that have substantial practical applications. I am very selective with the subject matter, and only teach my own courses. I never take over other people's courses."

▶    Do people do that – teach other people's courses?

"Yes, it's common practice. Taking over curricula and course materials. Some departments even have rules for that.. At the Physics Department, for example, if you develop a course you're only allowed to teach it for five years. Then someone else has to take over. But to return to the most important thing I teach my students – it's probably this meticulous accuracy – are we certain we are right? It forces us to go out and learn a mass of things. That's what I perceive as the purpose and major advantage of mathematics – to train people who can guarantee the journey from A to B. Otherwise we end up opinionating about everything under the sun. Just look at the blogosphere. People opinionate forcefully on everything all the time. We operate at the other extreme. We don't have much in the way of opinions. Someone gives us A, and then we do the rather tedious job of getting to B. I am a firm believer that we must educate people who can guarantee the quality of the journey from A to B. The world is obviously not so simple that it can always be done, but in very many engineering projects it is possible. For example, someone has calculated precisely how strong the walls in this building need to be to allow such gigantic windows."

▶    But you cannot mathematically calculate what thoughts are going on in here?

"No, we can't. Yet. By the way, I believe that we will be able to *measure* thoughts before we can explain them mathematically. The big philosophical question is whether mathematics, which is often used to predict the future, will ever be able to tell us what we are going to think."

▶    Which brings us to belief systems – what do you believe in?

"I believe in knowledge. I also believe that the human race is pretty destructive. Up through history there have been many wars and a lot of violence round about. And it's very easy to provoke men to violence. It's probably in our genes. The men who managed to bludgeon down the neighbouring tribe had more children."

▶    But is it still like that? Is violence an expedient behaviour for procreation?

"Yes, which brings us back to the question of smell. Perhaps the answer is that women are attracted to men with those genes. I'm afraid there's something in that."

▶    So we have not come so very far after all?

"No, and there's been quite a lot of research done on why women choose the wrong men. Perhaps it's hormonally predicated that wicked men should rule the world?"

▶   But is it right? According to Harvard professor Steven Pinker, the world has never been more peaceful than it is today. Perhaps it is little, lithe women like Aung San Suu Kyi who rule the world? In any case it is the outlook of Martin Luther King and Nelson Mandela that lives on and inspires us – not the mind-set of Adolf Hitler or Idi Amin. But if you really believe in your own pessimistic world view, I have to ask – bearing in mind Richard Hamming's philosophy of researching the big, important things – what do you, with your substantial capabilities, intend to do to change it? Or to put it another way: what is the discovery you would most like someone to make – for society as a whole?

"It would have to be that the good and kind knew with certainty that it pays to be good and kind. This brings us back to the educational reform we talked about at the start of this interview. The one which should have been implemented 30 years ago. Let's say that we have now established a different culture in academia, a different climate of cooperation, a more peaceful culture if you will. And it proves beneficial enough to outcompete the warrior culture that has dominated human history. The question then becomes whether such a cooperative culture is sustainable, or whether it will simply run aground. Cooperation is so demanding, particularly if the parties are dissimilar. Is it really a culture of cooperation that will drive the world forward? I'm afraid that laziness is the greatest enemy of kind cultures. If we get too comfortable, we quickly get lazy. The warrior has no choice but to struggle on."

▶   But what about Simula's own philosophy? Is it the thirst for conquest that drives the world forward, or happiness and curiosity?

"Simula's success is actually qualitative proof that cooperation, wonderment, curiosity and happiness are a superior strategy. In this way we have actually established as fact that it pays to be nice, by which I mean offering recognition and being generous, at the same time as this must be balanced by clear expectations and high ambitions. The highly competitive, self-promoting researcher is the metaphorical warrior, doomed to lose – in the long term. The good will probably win anyway. This is something that I can tell I need to think about some more. It almost seems as though a lot of my work at Simula and the UiO is actually contributing to the discovery that I most want someone to make."

And here we leave professor Hans Petter Langtangen. But it would not surprise me if one day he receives the Nobel Prize for discovering the mathematical formulas for feelings.

# Solving Puzzle Earth by Adaptive Mesh Refinement

## An Interview with Carsten Burstedde by Dana Mackenzie

At the first session of the Challenges in Computing Conference, Carsten Burstedde received the Springer Computational Science and Engineering (CSE) Prize for 2011. The prize is presented every two years to a team of researchers who have collaborated on an outstanding interdisciplinary project. At least two of the recipients must work in different fields of science and they must be less than 40 years old at the time of the award.

Burstedde was recognized for his work on a computational model of convection in the Earth's mantle, developed while he was a postdoc and then a staff researcher at the University of Texas. Burstedde has recently returned to Germany to accept a professorship at the University of Bonn, where he earned his doctorate in 2005. He shares the CSE Prize with Georg Stadler of the University of Texas and Laura Alisic of the California Institute of Technology.

The award-winning project merged Burstedde and Stadler's talents in parallel programming, finite element modelling and applied mathematics on supercomputers with Alisic's expertise in geophysics. Burstedde's most significant contribution was the development of new algorithms for adaptive mesh refinement that could be implemented on a supercomputer with as many as 200,000 processors.

Mantle convection is the process that is believed to drive plate tectonics. Over a period of hundreds of millions of years, the rocks in the Earth's mantle – which extends about 3000 kilometres below the surface – circulate in a rolling motion similar to the boiling of water in a pot. This motion causes the tectonic plates at the Earth's surface to shift, sometimes moving apart, sometimes rubbing against each other, and sometimes diving underneath one another in a process called subduction. The immense convection rolls span thousands of kilometres and yet some of the most important processes, such as subduction, earthquakes, and volcanic eruptions, occur on a scale of single kilometres. A truly accurate computer model of mantle convection, therefore, needs to be able to operate at both of these scales.

At present, even the fastest supercomputers cannot run a model at a one-kilometre resolution that spans the entire globe. Such a model would require a grid with more than a trillion cells, several trillion data points, and several trillion equations to solve. Not only is it impractical, it's unnecessary. The one-kilometre resolution is only needed in certain areas of the Earth's crust, for instance, the places where plates collide. In other areas, such as the interior of tectonic plates and lower mantle zones, a resolution of tens or hundreds of kilometres is good enough.

It is already a challenge to program a computer to solve the equations governing mantle convection on a uniform rectangular grid, but doing so on an adaptive grid whose cell size varies from one place to another increases the difficulty dramatically. In fact, even organizing the data takes serious thought. Computer scientists use a data structure called an octree, a branching network in which each node is either a leaf (an endpoint) or has eight nodes below it. The nodes represent cells of the adaptive mesh. Each time a cell is subdi-

A. Bruaset, A. Tveito (Eds.), *Conversations about Challenges in Computing*,
DOI 10.1007/978-3-319-00209-5_5, © Springer International Publishing Switzerland 2013

vided, it creates eight smaller cells, represented by eight new nodes of the octree. The finest cells in the mesh, which have not been subdivided yet, correspond to leaves of the octree.

Although adaptive mesh refinement is hard, in general, for parallel computers, Burstedde and other students in Ghattas' group have had good results with octrees and their two-dimensional cousins quadtrees. Each leaf needs to be apportioned to one processor in such a way that each processor does roughly the same amount of work. (Otherwise the parallelization will not be very efficient.) Managing the computation to avoid redundancy and minimize the time spent looking up data and communicating from one processor to another is a major challenge. One measure of Burstedde's success is the fact that only 0.05 percent of the computer's time was taken up on 'mesh operations' – essentially, bookkeeping – leaving 99.95 percent of the time for solving the physical equations. He also scaled up the algorithms from a few thousand cores to 200,000.

A final technical wrinkle Burstedde's team had to overcome is the fact that the Earth's mantle is not a cube; it is a hollow spherical shell. Their model divides the mantle into 24 cubelike regions, each one with its own adaptive mesh and its own octree of data. The software has to keep track of which leaves in different octrees actually correspond to adjacent parts of the Earth's mantle. Burstedde describes the resulting data structure as a 'forest of octrees', which accounts for the name of the library of algorithms for which he was the lead author: p4est.

The mantle simulation called Rhea was the first large team effort of Burstedde and Stadler, along with Lucas Wilcox. Eventually, Burstedde and Wilcox focused on the parallel adaptive mesh refinement code, while Stadler developed Rhea into a real mantle simulation tool that inputs temperature data and calculates velocities and pressures in mantle convection. In order to simulate the Earth, the octrees need to be populated with data, representing the physical properties (temperature, viscosity) of each of the corresponding cells; then the equations of motion have to be solved (giving the velocity field in each cell) and finally the solutions need to be compared with reality.

Once the software library was in place, then the interdisciplinary fun could begin. Inevitably the model failed to match reality at first; that is because many of the physical parameters that went into the model are actually unknown. The modellers continued to adjust those parameters until the results of the simulation agreed with observations to a reasonable extent. Once a reasonable match was found, the team inferred that they had found the right values of the hidden parameters. Now the model could be used to answer geological and geophysical questions. For instance, in a paper published in *Science,* the team reported that a much lower amount of energy than expected was dissipated in 'hinge zones', and a larger amount of energy in the lower mantle.

The model can, in theory, be run backwards and forwards in time to find out tectonic events that happened millions of years ago or forecast events millions of years in the future. However, Burstedde said that the team has not attempted anything so ambitious yet: 'Just to reproduce present-day plate motion made us incredibly happy'.

The senior members of the research team (who were not eligible for the CSE prize, due to the age restriction) were Omar Ghattas from the University of Texas and Michael Gurnis from Caltech. Lucas Wilcox, who is now at the Naval Postgraduate School in Monterey, California, also contributed to the project, as mentioned in the interview below.

The interview was conducted on December 28, 2011.

▶    How did you end up at the University of Texas?

I finished my PhD in Bonn in 2005. That year, in September, I had already found this advertisement by Omar who was looking for postdocs at the University of Texas. I contacted him and thought about it for a while and in the spring I accepted. I showed up in May of 2006 as a postdoc.

▶    Did you know Dr. Ghattas at that point?

I met him at a conference in Europe in 2003 and I talked briefly with him. I was honestly very surprised that he remembered me when I e-mailed him years later. Now I know that

Omar remembers everybody, but at the time I was kind of surprised. What entered into my decision the most was a phone call we had, where we talked about the position for 40 minutes or so.

▶     Did he tell you at that time about this idea of modelling the mantle?

Not yet, that came into picture more in 2007. He was mostly talking about the potential in supercomputing. He had started in Texas not too long ago at that time and he was already connected to the Texas Advanced Computing Center (TACC), so he realized the potential that was there. Of course, he was right in predicting that Texas would become a very important player in the supercomputing scene.

▶     Did you have any experience of supercomputing before?

Not at all.

▶     What made you think you could do it?

I didn't mind starting something fairly new. I did the same kind of thing when I worked on my PhD, switching from physics to numerical methods, with no in-depth knowledge of that area. I felt fairly good about it.

▶     How much time pressure was there? Did you have to submit grant proposals right away?

Omar handled most of the grant writing himself, with input from us. It was important for us to be part of the competition at the TeraGrid 2008 conference in Las Vegas. We won the TeraGrid Capability Computing Challenge award. Two months after that was the deadline for the SC Conference (formerly called the Supercomputing Conference), one of the biggest in this area. The Gordon Bell Prize, one of the more prestigious supercomputing awards, is awarded during that conference and actually Omar had won it before in 2003. We submitted for this award both in 2008 and 2010. We were finalists both times but we never won.

▶     Not every field of science is so award oriented. Is that a particular feature of the supercomputing area?

It seems to be a fairly important social component of this whole academic life. It wasn't clear to me in the beginning why we should submit for these competitions. Just working on the project was fun enough; we didn't do it just for the award. Of course, it was a great experience to see how this whole SC Conference is set up and how it works.

▶     Can you tell me how this particular project got started?

Omar and Mike got together and submitted a grant proposal to the National Science Foundation. In 2006, when I got to Austin, this whole corner of the building was basically empty. I showed up in May, Lucas in August, and Georg in September. That was basically the core of the group for the next couple of years. I started getting the adaptive mesh refinement (AMR) framework going.

In the beginning there was an AMR library called OCTOR written by a former student of Omar at Carnegie Mellon, Tiankai 'TK' Tu. That was working on the box, but it was rather scalable and it was the basis of the work that won the Gordon Bell Prize in 2003. We wanted to create a layer that could use this and put our own discretization on top of it. Lucas, TK, and I designed it and I found it so much fun that I started programming much of this layer. I put in some parallel vector logic and basic Laplace operators, matrix–vector products, things like that. I talked a lot with Lucas, who contributed some important code as well. Georg put in the time evolution so that we could actually run the system forwards in time. That was the first solver for the convection–diffusion equations. If you have a given velocity field, you have a heat distribution that advects with the velocity, but at the same time it's diffusing independently of the current. That was the first work we did that was presentable, which won the 2008 TeraGrid award.

▶   What do you mean when you say that OCTOR was working on the box?

It means the domain is basically a cube. You can always cut any cube into eight smaller cubes, by cutting it once horizontally and vertically and back to front. You can do the same thing recursively again. But for the sphere, for the Earth, we needed a more complicated geometric environment. OCTOR did not support computational domains that are not mappable to a cube.

I always want to do my own thing when it comes to coding, so I started to derive this new parallel AMR code on more general geometry. In the end it became available as the p4est library. In the beginning I talked a lot with Lucas, who contributed important code as well. Parallel to me embarking on general geometry AMR, Georg embarked on adding the Stokes operator and wrote the finite element code that would make it work on the sphere.

▶   What is the Stokes operator?

The Stokes operator is the one that specifies viscous flow. When I was talking about advection earlier, that was advancing information along streamlines without any forces acting between streamlines. But in viscous flow, if you imagine honey or lava, the velocity is not just given so you can advance the temperature along the streamlines. In this case the velocity is the unknown. If you squeeze honey and it starts moving, the velocity field is a result of applying forces to this viscous fluid. The Stokes operator defines how the velocity field interacts with the resisting forces. The velocity field has a direction, so it is a three-dimensional unknown, as opposed to just a one-dimensional unknown such as temperature. So viscous flow gives a vector-valued system with a lot more structure. Unlike the advection equation, the solution of the Stokes operator system requires solving a very large nonlinear system of equations with up to a few billion unknowns.

From then on, that was when Mike and Laura really came in the picture. Once we had the spherical Stokes solver, we wanted to know what kind of temperature background fields we needed and what kind of rheology parameters. That's when all this shuffling back and forth happened. We would use some of this data, visualize the solutions, and send them back to Mike and ask how that looked. Later we showed him how he could himself use the visualization capabilities. We would use this remote visualization cluster at TACC called Spur. We could run something on [the supercomputer] Ranger, leave the data there, and then Mike would run the remote visualization on the same system and look at the data himself. Georg ran more runs with different data until we converged on what we wanted. The main interaction was in the preparation of the input data and the quantitative and visual interpretation of the results. Since Mike and Laura were in California, we used weekly Skype teleconferences to discuss that status of our collaboration.

▶   Can you describe what adjustments you would make when he saw one of these images that didn't seem to match? What could you adjust?

We had a couple of rheology parameters that were global, where tweaking one number would affect every place on the planet. Those are crude estimates, because we knew there is no way the Earth could be that symmetric. But we didn't have much of a choice because we had so many parameters to fiddle with.

Then there were also the input data sets that were specifying slabs and subduction zones. Mike had observations where fracture zones are specified on the surface of the planet. Geologists knew what angle they would be diving into the depths at, say, 30 degrees. We had to extrude this information into the three-dimensional mantle, to transport our knowledge of the temperatures along those features. There were a whole bunch of iterations, tuning those geometric features. Imagine some wicked, curved line that's some kind of boundary between plates which at some depth starts folding over itself or rolling over like a pretzel. Again, visualization was the key to figuring out how good those geometric features were. Along with the geometric features, some numbers came along that said how weak the zones were. We could tune those on a case-by-case basis. One of the worst places was along the west coast of South America, the Nazca plate, a small plate wedged between the Pacific and Andes. That thing was just out of whack before we fixed it. It's still not great, but it's a lot better now.

▶  Does the difficulty of that particular location reflect something in the geo-physics that's hard to model?

Probably yes. We're just really crude still because we have such globally specified parameter values. This leads us into the future research. With a couple people who share a visualization channel and a computational resource, we can run a hundred simulations a year, using our personal time to interpret the data and improve the model. But that is if we have only five parameters for rheology and those line maps on the surface. Now imagine we have a different parameter in every subduction zone or a different parameter at every depth level. The possibilities we have to choose will just explode and we will never have time to play with it sufficiently. This is the point where we need mathematical methods to help us choose those parameters automatically.

What this leads to mathematically is inverse problems. We can compare the difference between our outcome and the real motions everywhere and that gives us an error value. This error value is a function of the set of all the parameters we have. We want to minimize this function just as in high school calculus. The whole simulation is like a giant side condition to this minimization problem. Whenever we change parameters, we have to solve the whole mantle system for the velocity.

That is basically the idea behind the inverse problem. In principle, we know how to do it. Omar has a lot of expertise doing this in the context of seismology and Georg and I have worked with it in different contexts. This is one way that we can advance the model at the level of local detail, without doing all of this matching manually.

▶  Would you say that the biggest challenges are algorithmic or mathemati-cal or with integrating the algorithm with the science?

We want to improve both the discretization and the solver so that we can go to finer scales, in both space and time. We're curious about some of the microplate systems. At some point we will have to accommodate additional physical modelling that is not relevant at the cur-rent scale but at finer scales will be important. For example, Laura is planning to work on mantle–magma interaction. That is a finer-scale physical process. There is more liquid than we allow in our current models and geochemistry and possibly other kinds of mechanics. If we want to integrate even finer scales, we will really be challenged in terms of the modelling. Of course, there will be another explosion in the number of parameters, which gives us an-other reason to work on inverse problems, to make this practical or even possible.

For now we are doing groundwork on time evolution. There was a question asked at the conference about going 10 million years ahead or 80 million years back. That's some-thing we are looking at in the short term. The real multiphysics stuff is far ahead.

▶  We've so far been talking mostly about mantle convection work. Are there other applications you're working on?

Our second big application is seismic wave propagation, that is, simulating earthquakes. Again, that's something we started from scratch, mostly driven by Lucas because he had interest in high-order wave propagation methods. He had a large part in the seismic wave propagation code, which works well now in all sorts of toy problems. But in this project we're not quite as advanced in terms of specifying realistic physical situations, like a full Earth. We are still ramping up collaboration with geoscientists, who will tell us where we should go in terms of scenarios.

Then, of course, I'm interested in AMR as a technology and in how it will work in future environments and architectures, irrespective of the concrete application. Right now AMR is still an exception; most people who want to run a simulation start out on a regular grid. It would be nice if in the future AMR were a default that everyone had already which doesn't cost very much.

▶    **Why is AMR not yet a default method? Are there some technical difficulties in implementing it?**

In parallel computation it was not clear that it would ever work. Most people, up to a couple years ago, were quite sceptical. Now there are a couple other approaches that are also scaling fairly reasonably.

At least in terms of mesh sampling, we have discretization issues. There are people who don't like working with T intersections, or dangling nodes, which you get if you imagine the adaptive grid with a large box on the left and two small boxes at the right. You need to implement a little bit of extra logic to make sure that you don't get holes or artefacts in the function you want to represent at a T junction. We take measures to enforce continuity, but still some people are not convinced that they will not introduce spurious features in simulations. It's just a matter of opinion now. Also, you need to take this burden away from the implementer who just wants to try out a new physical system. You shouldn't expect all this extra logic from someone who just wants to play with it.

There is algorithmic work that needs to be done, to make sure that our mathematical algorithms achieve a result without wasting resources more than necessary. It only makes sense to do this before we think about putting the problem on a bigger machine, with more CPUs or processing cores than previously. But suppose we are at the point where we necessarily have to think big. I think that handling this higher amount of concurrency can be done in the context of AMR. I don't see a bottleneck in principle there.

What is harder to judge is the fact that the memory access speed is not growing nearly as fast as the amount of processing force. Even now simulations are starved by not getting enough data, so they are not able to compute, even if the processors are idle because the data are just not there yet. This is related to the memory bandwidth, the rate at which data are coming out of memory. The problem is getting a lot worse because memory bandwidth is just not growing any more.

It's likely we will have to go back to develop new algorithms that are more favourable in terms of memory requirements. I think that predicting 10 years ahead is really hard. Even five years from now, I'm not exactly sure what we will be doing about the memory bottleneck.

▶    **What do you think are the most important ingredients that enabled your group to be the first to simulate the mantle down to the kilometre scale?**

We managed to generate a lot of synergy. First, there was a synergy between Georg and Lucas and me, coming in with different backgrounds and different preferences of what we like to work on and what we consider not so important or not so fun. Then we had synergy with the geophysicists, Mike and Laura, who were able to understand most of our problems in simulating those equations. Also, thanks to Mike's prior experience in parallel computing, they knew enough that we could exchange data and talk about what we wanted to measure.

▶    **What did you enjoy most about this project?**

I had a lot of fun working on the AMR algorithms because I like the geometrical problem setting. First of all, there is the geometry of the Earth, then the geometry in the equations, and the three-dimensional nature of the Stokes equations that relate the strain and stress. You can make drawings to see how the stresses interact with the strains. There is also geometry in the mesh and in the connections between different corners and intersections and lines and boxes. Of course, it is wonderful to work with the real Earth and not just toy problems. It was a great project that made me very happy to work on it and I'm glad that the results turned out so well.

# Computational Inverse Problems Can Drive a Big Data Revolution

*An Interview with Omar Ghattas by Dana Mackenzie*

The attendees at Simula's Challenges in Computing conference were privileged to receive a double dose of geophysical science. First, Carsten Burstedde was named a co-winner of the Springer Computational Science and Engineering (CSE) Award for his work on mantle convection simulation. In addition, his mentor, Omar Ghattas of the University of Texas, was one of the eight invited speakers at the meeting. While Burstedde lectured on adaptive mesh refinement in simulations of the Earth's mantle flow, Ghattas cast his net more broadly and outlined five 'grand challenges' in scientific computing.

Geoscience is currently undergoing a 'perfect storm', as Ghattas described it: a convergence of immense amounts of sensor data, new supercomputers to analyse it, and improved mathematical models to plug the data into. All of these converging streams have to funnel through a bottleneck known as inverse problems. Without fundamental improvements in this essentially computational problem, Ghattas argued, we will lose much of the opportunity for extracting geophysical knowledge from the data deluge.

For example, the field of seismic imaging currently makes use of only a tiny fraction of the available data. A National Science Foundation (NSF) project called USArray is setting up a grid of seismic stations at 70-kilometre intervals across the country. (Several other countries have similar projects.) Whenever an earthquake takes place, each seismometer records the resulting ground motion. The shape of the recorded signal is determined by the elastic properties of the rock through which the seismic wave passes. In conventional seismic tomography, most of this information is discarded, retaining only a few key numbers: the arrival times of different kinds of waves (S-waves, P-waves, etc.) As a result, these numbers do not provide the full story about the interior of the Earth.

'You can see that for stations close to each other, the arrival times are essentially the same', Ghattas said in his lecture. 'But they can have very different wave signatures that reflect the structure of the Earth underneath. We'd like to use the full wave equation, taking data from all of these instruments around the world and integrating them through an inverse problem to solve for the distribution of elastic properties inside the Earth. There's a race to do it now'.

In a 'forward' problem, often represented by a set of partial differential equations, a scientist assumes that the properties of the physical system being studied are known. In geophysics, that would mean that the scientist assumes a specific model for the layering of rock beneath the Earth's surface and plugs in known values for such parameters as the density and stiffness of the rock. The geoscientist can then run a simulation *forwards*, from cause to effect, to answer questions such as if an earthquake happens at this particular point on the San Andreas Fault, what will be the effect on Los Angeles?

Inverse problems, on the other hand, go *backwards*, from effects to causes. A train of waves arrives at a seismic station. What caused those waves? The simplest question – where the earthquake occurred – can be answered using arrival times and triangula-

A. Bruaset, A. Tveito (Eds.), *Conversations about Challenges in Computing*,
DOI 10.1007/978-3-319-00209-5_6, © Springer International Publishing Switzerland 2013

tion. But the more difficult question concerns what kind of material those waves passed through. And that is what scientists often want to find out, because it tells them the parameters they need to populate the forward model described above. How are they ever going to find out the properties of rock hundreds of kilometres underground? They can't put instruments there (and won't be able to in the foreseeable future), so the only way is to solve the inverse problem.

The most obvious and least efficient way to solve an inverse problem is by guessing. Try one set of parameter values, solve the forward problem, and see if it matches the observed data. If not, keep trying. Of course, mathematicians have devised vastly better algorithms than that which involve inverting large matrices. And yet the entries in these matrices … come from forward simulations. And lots of them. If you want to solve an inverse problem with 100,000 parameters (corresponding to the stiffness at 100,000 points in the Earth), you have to solve the forward problem about 500,000 times. 'That is just out of the question', Ghattas said. And yet 100,000 parameters are not that many! This corresponds to all the points in a $10 \times 100 \times 100$ grid, which is very coarse if you are trying to model the whole Earth. 'We need new thinking, brand new ways of looking at this problem', Ghattas concluded. He suggested that probabilistic algorithms might be one possibility.

Two of Ghattas's grand challenges concerned forward problems: to improve adaptive mesh refinement algorithms and to develop scalable parallel solvers. He noted that partial differential equation (PDE) solvers unfortunately generally run rather inefficiently on massive parallel computers. 'It's like owning a Ferrari that can go 175 mph and living in the city and never going faster than 25 mph', Ghattas said. 'Maybe these high-performance computers are not optimally suited for PDE problems and we shouldn't worry about getting a high percentage of the peak performance. Yet government agencies are funding these machines and they don't want to hear that you're getting only 7 per cent of the peak'. Clearly, improvements in forward solvers will be necessary as long as we have to solve thousands of forward problems for every one inverse problem.

Ghattas's last grand challenge was to educate students who will be capable of solving these problems. 'Universities have fallen behind in providing students with the knowledge to design algorithms and write software for modern architectures', he said, quoting an NSF task force he had recently served on. He is concerned that computational science and engineering as historically fallen into the cracks between disciplines.

The following interview took place on February 22, 2012.

▶    **Could you tell me about your educational background? In particular, how did you arrive at this unusual combination of interests, geoscience and computational science?**

I had a cousin who is a civil engineer and worked for a construction company in Qatar. My father worked for Aramco and we lived in Saudi Arabia, in a company town. My cousin would take me out to construction sites and I was always excited by these massive projects. So in the fifth grade I already wanted to be a civil engineer.

I went to college at Duke University, where I majored in engineering and took the standard complement of undergraduate courses. I decided to stay on as a graduate student in order to go deeper into subject. Civil engineering at the undergraduate level contains many subspecialties. You study subsurface soils and the response of structures to loading. There's environmental engineering, coastal, and wastewater. I was interested in structural mechanics and went into grad school in that area. Duke is a relatively small university, and in small departments you quickly exhaust all the classes in your area. They encouraged classes outside the department in computer science and mathematics, so I took numerical analysis and applied mathematics. I began to appreciate that the future was at the intersection of those three subjects: computer science, applied mathematics, and continuum mechanics. The interface is computational mechanics.

▶    **How did you decide to go to Carnegie Mellon and what sort of research did you do there?**

After I got my degree at Duke, Carnegie Mellon (CMU) was advertising for a position in computational mechanics. I don't think I applied to more than three faculty positions when I was coming out of graduate school. I was fortunate to get an offer there and accepted.

The nice thing about CMU is that it's a small university, on the order of 5000 or so students. Because of that small size, there was a very high degree of interaction and inter-disciplinarity. At CMU the notion of interdisciplinary research was not just talked about but actively encouraged. People and research groups would regularly cross college boundaries. It was very much part of the DNA of the university. I was in a department that was not just open to the idea of computational mechanics but also identified it as a future growth area.

Also, being at a university in which we had a top-ranked computer science dept., computing was a big deal at CMU and it cut across most academic programs. In 1989, when I went there, we had the Pittsburgh Supercomputing Center, one of four in the nation along with the National Center for Supercomputing Applications, San Diego, and Cornell. Pittsburgh always had a long history of getting the latest machines from Cray. They also had some experimental machines from Thinking Machines Corporation, the CM-2, the first massively parallel computer. These computers would seriously challenge the way we thought about algorithms. In distributed memory machines, each processor has its own memory. You have to explicitly think about how to partition the problem into pieces and worry about how all these processors would interact.

At the time I began thinking about the algorithms that are the workhorses of scientific computing. How do we redesign them to scale to 512 processors by the mid-1990s, a thousand by the late 1990s, and now hundreds of thousands?

▶   This all still sounds pretty far removed from geoscience. How did that begin to enter the picture?

One of my collaborators at CMU was Jacobo Bielak, an expert in earthquake engineering and engineering seismology. In the early 1990s he had this vision that we could model the propagation of seismic waves in the Earth in three dimensions. We could model an earthquake and study its influence on ground motion, using these supercomputers. We thought that perhaps now was the right time; we finally had machines powerful enough to do this.

That began a long collaboration on modelling the problem directly. We would model the rupture itself, then the wave propagation, and then how the ground motion would arise in a sedimentary basin such as the Los Angeles Basin, with soft sediment surrounded by mountains. Waves tend to be amplified in the basin. If your house is built on a surrounding hillside, you might experience an order of magnitude less shaking than you would on the soft sediment. You really can't tell this just by looking at the properties of the soil beneath you. You need a three-dimensional model of how waves propagate through the Earth and get amplified by the complex structure of the basin.

At this point no large-scale calculations had been done on this problem. What it amounts to is solving the elastic wave equations, so you have both shear waves and longitudinal waves. The challenge is one of length scales. There is significant earthquake energy up to several hertz, which is considered a low frequency for exploration geophysics but a very high frequency in this business. A characteristic shear wave travels at speeds of 100 metres per second or less in typical soft sediments. For a 1-hertz wave, you have a 100-metre wavelength. If you're numerically solving the equations, you need something like 10-metre grid spacing. If you're modelling a rupture on the San Andreas Fault, your mesh has to stretch from north of Los Angeles to south of San Diego and from east of the San Andreas to the Pacific Ocean. This is a volume 800 kilometres long, 100 kilometres wide, and 100 kilometres deep. You can't put a 10-metre grid spacing on it.

So we developed adaptive resolution methods based on so-called octrees. You saw them mentioned in Carsten Burstedde's talk. We concentrate the grid resolution in sediments where the wavelengths are the shortest. Outside of the sedimentary regions you can have much coarser grids, where wave speed is 6 or 8 kilometres per second. That, combined with the emergence of parallel machines, was great news because it offered the possibility of enough speed to do these earthquake simulations.

Even with the raw performance available, we still needed to develop advanced algorithms and good parallel implementation of these algorithms so that the load would be uniformly distributed and we wouldn't have processors sitting idle. Also we needed was to reduce the amount of communication across processors. That's the Achilles heel of parallel processing. If you spend too much time communicating, you lose the advantage of having so many processors. These issues forced us to rethink a lot of the algorithms we used. Within a few years we were running models of ground shaking at frequencies of engineering interest and we got up to 2 hertz by the time I left.

Let me clarify that we're not trying to predict when an earthquake will happen. The process of rupture and what leads to it is a very difficult problem, with scales ranging from millimetres up to hundreds of kilometres. It's an unresolved question how to model that. Instead, what we were doing was to postulate an earthquake, to suppose, for example, that a rupture occurs on a certain fault in a certain way and with a certain amount of energy. By considering a collection of such probable rupture scenarios, we could ask what the consequences are in terms of ground shaking and resulting structural damage for the Los Angeles region. That would amount to putting a known source into the elastic wave equations and solving them repeatedly, an extremely difficult problem but a forward problem.

▶    When did you start getting interested in the inverse problems? Can you describe the difficulties you faced?

I really got interested in them in the late 90s while at CMU and extending into my time here at Texas.

▶    When did you move to Texas and why?

In 2005 I was invited to move to the Institute for Computational Engineering and Sciences (ICES), an interdisciplinary institute that brings in professors from all around the university. It doesn't sit within any particular college. Its mission is to address problems of great scientific and societal importance through mathematical and computational modelling. We have 80 faculty and well over 70 PhD students now. At the time I moved here, I also joined the Geological Sciences and Mechanical Engineering departments. All of the faculty in the institute have home departments, but they work at the institute and do research at the interfaces of established disciplines.

ICES's leadership in modelling and simulation as well as the strengths of the University of Texas' engineering and science programs brought me to Texas. Also, of course, Austin is a nice town. In the past, we used to address these complex and difficult problems from the perspective of one discipline. Increasingly, that's not a tenable approach to the problems that face society. Every branch of science and engineering has been impacted by modelling and simulation, from climate change to the design of medical devices and the planning of surgical procedures.

▶    Getting back to inverse problems, could you describe how they emerge in geoscience and what accounts for the difficulty of these problems?

In a seismic inverse problem we don't really know the mechanical properties of the rock, the density, and the elastic moduli. Those can be roughly estimated from geological models and you can get some information from drilling. But the actual heterogeneity of these properties within the basin is critical. It plays a role in how the waves propagate and how they transmit through interfaces or are reflected.

We have seismograms from past earthquakes whose sources are well characterized, particularly for small quakes. You only need the angle of inclination of the fault, the magnitude and evolution of the rupture, and the source location. From this information, can we solve the inverse problem, where we go backwards and try to figure out the mechanical properties of the rock, which amount to the elastic moduli and density in wave equations? Can we estimate the P- and S-wave velocities? This is a highly nonlinear inverse problem. It looks innocuous. The wave equation is linear in the elastic wave motion field (i.e., displacement or velocity at each point in the Earth). But the inverse problem is highly nonlinear.

We take what's known as an output least squares approach. We ask what the mechanical properties need to be so that the output (predicted wave motion) matches the observations with the least possible discrepancy for each earthquake under consideration. We're minimizing the $L_2$ norm in terms of the difference between the solution to the wave equation and the observed readings, summed over all seismometers. We are minimizing with respect to the Earth's parameters, which are coefficients of the wave equation. It's the inverse of the forward operator, with heterogeneous parameters inside of it. When you write the displacement as a function of the elastic coefficients, that function is highly nonlinear because of the heterogeneity.

This is a complex problem with lots of local minima. There's an even more insidious feature that it shares with many other inverse problems, in that it's ill posed. When you have a heterogeneous medium, the parameter you're inverting for is really a field: wave speed as a function of space. The data we are inverting with is band limited. That means we have information perhaps up to 1 hertz or 2 hertz if we're lucky. If I try to go backwards and try to infer Earth properties from this, I'm not going to be able to infer small length-scale features below, say, a quarter of a wavelength. There will be many modes, many wave-number components of the parameter field that the data are not informative about.

So we have this large null space consisting of Earth properties that are not uniquely identified by the data. In order to suppress these modes, we do regularization. We construct a penalty term that we add to the data misfit, which penalizes the highly oscillatory modes of the Earth model, the high wave-number part. We try to allow the data to determine the smoother components and let the regularity kill the high wave-number components.

So far we've made pretty good progress. Let me distinguish several different physical problems where seismic inversion is the crucial computational approach. One problem is at the regional scale, identifying sedimentary basin structure from past earthquakes. Although that is what I started out on, I haven't worked on that problem for a few years. A second level is the global seismology problem, where we try to determine the structure of the entire globe using the available seismic data. The Holy Grail is still to use data up to 1 hertz, where seismometers are capturing significant energy.

We do have existing models of the entire globe, which describe how wave speeds vary in the Earth. The basic model is one of radial variation, with concentric shells in which the wave speed varies linearly. But we know there are small lateral variations. These are caused by the transport of heat within the Earth. When you have plumes of warmer rock, it leads to a smaller wave speed. Also, there are variations in chemical composition. So the question is can we identify lateral variations in the wave speed? Our existing picture relies on travel times derived from simplified models of wave propagation, such as ray tracing. No one yet has been able to solve the problem at a global scale using a mathematical model based on the elastic wave equation. We're working on that right now and other groups are working on it.

Even the forward model alone is a challenge for supercomputing, to go up to 1 hertz at the global scale. We'll need the next generation of petascale computers. For the inverse problem, we'll need exascale. We expect that new structure will be revealed, particularly in the crust, by using all this new data.

> In your talk you mentioned probabilistic methods. Can you say more about
  how these can be used?

Everything I've been talking about so far is the classical or deterministic model. Find the solution that is closest to the observation. Predicted waveforms are connected to the parameters through the solution of the wave equation. That will give you a point solution, the best fit.

A few years ago I started getting interested in the statistical interpretation, in the framework of Bayesian inference. I was influenced by a book by Alberto Tarantola, who got into this 25 years ago. As I said earlier, there are many Earth models consistent with the data. The regularization scheme penalizes a whole family of Earth models in order to come up with the one that is most regular – in essence, it is Occam's razor.

Instead of Occam's razor, the Bayesian approach gives you a complete statistical description of *all* Earth models that are consistent with the data within a certain noise level.

It gives you the probability of each one. There's just one problem. The result, the solution of this Bayesian inverse problem, is a probability density function (pdf) in as many dimensions as there are parameters in the problem.

Say I discretize the Earth down to a 1-kilometre resolution. There are about 1 trillion cubic kilometres, so I have 1 trillion parameters. The techniques of today are woefully inadequate to describe this probability distribution, but this is my interest. The natural setting for many of these geophysical inverse problems, I've come to be convinced, is this statistical one. The natural solution should be a pdf. So we need radical improvements in our ability to explore the pdfs.

Markov chain Monte Carlo (MCMC) methods have become very popular for sampling multidimensional probability densities. We might be interested in computing the mean and the higher moments of the probability distribution. MCMC allows you to draw samples from these distributions. Even so, you need millions of sample points, even in a modest number of dimensions. That means you need at least as many solutions of the forward model. There's no way you can do it millions of times.

In our current research, we bridge the deterministic and statistical approaches. We start from Newton's method, in which one makes a local approximation of the surface we're minimizing, and try to bring these approximations into MCMC to accelerate the sampling process. In a sense we are using the methods for deterministic problems to endow the MCMC with knowledge of this surface in order to sample it more effectively.

I believe that one of the keys is going to be recognizing that even if you're in a million-dimensional space, the data won't be informative of more than a thousand of these. The rest of the information has to come from prior knowledge. If we can discover this low-dimensional manifold and exploit its structure, that's the only part where the expensive forward model has to get involved and we have a chance of overcoming the curse of dimensionality.

This is what we're working on now. In the next couple years, I think that breakthroughs will occur. It was unfortunate that I wasn't able to talk about this whole area of uncertainty quantification in inverse problems when I gave my talk in Oslo. I was trying to go slowly over the earlier stuff and when I got to the end, I didn't have time to talk about uncertainty quantification.

▶    If you could perform this full-wave inversion, what would you do with it? Give me some examples.

There are several different seismic inverse problems where the mathematical structure is the same. There's the regional setting, where you're trying to figure out the properties of, say, the Los Angeles Basin. If you were able to do that, you'd have greater confidence that when you run these forward models, the results are realistic. That will lead to more accurate predictions of the level of ground shaking. That's exactly the information that a structural engineer needs to know when he's designing earthquake-proof buildings: What is the horizontal acceleration?

Then there's the global seismology problem. At that scale, the prize is understanding the structure of the Earth and unlocking the secrets of geodynamics.

A third class of seismic inverse problems where the properties of the Earth are critical is in exploration geophysics – typically, exploring for oil. The oil companies solve inverse problems on a daily basis to try to predict where oil is. That is not being done routinely with the full wave equation, but with simplified models. The oil companies are all interested in using the full wave equation.

In all of these applications, except for the second, you will make some decision based on the structure of Earth. You want to guide the building codes for a city or you want to decide where to drill a well. All these decisions have to be made in the context of uncertainty.

That's why developing techniques for quantifying uncertainties will be so important. You want to say with a certain probability that this will be successful.

► In your talk you also mentioned that parallel computers are not very good at solving the forward differential equations. Why not?

This is really critical, an important problem. We are in the midst of a big revolution in high-end computing. We're seeing a transition from older computers with a single core, 10 years ago, to modern computers that have multiple cores and, through graphics processing units (GPUs), with hundreds and soon thousands of cores on a processor.

What's driving all of this is power budgets. Ten or 15 years ago each new processor would run at a higher clock rate to give you more flops per second. But then we hit a few gigahertz and flattened out. You could not just continue to increase processor speeds in an unlimited way. You just generate too much heat with increases in clock speed – there is too much heat dissipated and too much power consumed. Unfortunately, the scientific community and the general public are accustomed to a doubling of performance in each generation. The way to do that is to put more cores in a processor.

What is the implication for scientific algorithms? We're putting more cores in, but we're not increasing memory bandwidth in a consistent way. Much of the consumer market doesn't demand the high memory bandwidth of the scientific market.

► What is memory bandwidth?

The speed at which you can pull data out of memory into processors. Much of computational science involves sparse calculations, for example, a PDE that you discretize with finite element methods. You're doing an update at one grid point that only involves nearby grid points. The problem is when you do an operation, you have very few flops before you have to get new data from memory. If you can grab new data from memory as fast as you can do a flop, then you're fine. In older processors, they were much closer to each other. But now you can do hundreds of flops for the cost of going back to the main memory. So we have machines that are not as optimally suited to scientific computation. Things are going to get much worse when we go to systems with hundreds of cores. Unless memory bandwidth increases consistently, we won't be able to make effective use of them.

Denser matrix operations will benefit more from high-speed computing, because you have many more flops than memory accesses. Part of the challenge is whether we can reconfigure the algorithms to make them rich in dense matrix computations. Some people would say that we should go to higher-order methods. Why approximate behaviour as linear? Why not have a sixth- or eighth-degree polynomial representing each element?

In my talk I showed a table where we're scaling up to 220,000 processor cores on the Jaguar supercomputer at Oak Ridge National Laboratory, with 99 percent efficiency, using sixth-order elements. For wave propagation, having a high order is a good thing because you get phase errors that accumulate over time. They're best addressed by a high-order method. Not all scientific computing problems have that possibility.

One area that is a challenge is implicit solvers. These are wave propagation methods where we never have to invert a matrix, we simply do matrix–vector multiplications. These methods are critical for problems with a wide range of time scales, where small scales are not important but are present in the numerical model. An implicit solver can step over them without becoming unstable. I think I showed some examples in my talk when I spoke about the dynamics of ice sheets. There implicit solvers are critical. Unfortunately, it's not clear at all how to get them to work well with the new generation of many-core processors.

► For those of us who don't keep up with the progress of supercomputers, how big and how fast are the current generation of machines?

Jaguar at Oak Ridge now has 300,000 cores and it is still the largest supercomputer available for open science research in the US. It currently runs at 3 petaflops (a quadrillion operations per second). Within a year it's being upgraded to 10 petaflops and eventually it is planned to reach 20.

Here at Texas we're deploying a 10-petaflop computer. The Texas Advanced Computing Center will host this new machine, called Stampede, which is supported by the NSF. Eighty percent of those 10 petaflops are coming from a many-core processor from Intel, called Many Integrated Core (MIC). We're all thinking of ways to exploit these processors.

▶    At the end of your talk, you mentioned education as one of your grand challenges for computer science and engineering. Could you elaborate on that?

Today's problems involve more complex mathematical models that require more sophisticated computer science techniques. In the old days you could focus on deterministic solutions of single-physics problems in forward mode. Now we are trying to solve inverse problems in a stochastic setting for multiscale and multiphysics problems. Students will need a strong background in mathematics. Also, computer architectures are continuing to become more complicated, with GPUs and memory hierarchy and multiple processors. These too require special training.

Finally, it is important for students to be trained in an interdisciplinary environment that blends mathematics and computer science with an application area. At ICES the students take courses, have to pass qualifying exams, and do their research in all three areas.

My sense is that in Europe, Japan, and China the recognition of the importance of educational and research programs in CSE is further along and there are more programs in place. The World Technology Evaluation Center evaluated research programs around the world and said that the US is falling behind. I don't want to say that there is only one solution – different universities will do it in different ways. But this is a critical issue and we need new kinds of educational programs.

# Towards the 'Google Heart'

## An Interview with Natalia Trayanova by Dana Mackenzie

Heart disease is a life or death issue for millions of people. In the United States, it is the leading cause of death every year and yet many of the treatments for heart disease are still extraordinarily crude. For example, the primary treatment for ventricular fibrillation, a lethal heart rhythm disorder, is to give a huge electric shock to the heart. 'It's as if you want to open the door and you don't have a key, so you blow up the door instead', says Natalia Trayanova, a physicist and biomedical engineer who develops computer models of the heart.

The price for our lack of understanding of the mechanisms of heart disease is considerable. Even though defibrillators save lives, they can cause psychological trauma for people who undergo the shocks repeatedly (e.g., patients with implanted devices). Likewise, drugs that fail because we don't understand their mechanism of action cost money to pharmaceutical companies and leave people with rhythm disorders at risk for later heart attacks.

Trayanova envisions using computer models as a 'Google Heart' that will allow doctors to understand what is happening in their patients in the same way that Google Earth gives geologists a bird's-eye view of our planet. 'I would like to bring computer simulation of the heart function to the bedside', Trayanova says. (Actually, she says, in the office of the treating cardiologist would be good enough.)

At the Challenges in Computing conference, Trayanova spoke about the cardiac models that she has developed over several years at Tulane University and Johns Hopkins University. Those models have been moving closer to clinical application and towards Trayanova's ultimate dream of using computer simulations to tailor cardiac therapy to the specific patient.

Trayanova discussed her team's on-going work on catheter ablation, a technique used to treat ventricular tachycardia in patients who have had myocardial infarction. This is a life-threatening arrhythmia. The catheter ablation procedure takes a long time – four to 12 hours – and it is still a hit or miss proposition. Cardiac electrophysiologists have to stimulate different parts of the heart and try to identify which regions are causing and sustaining the arrhythmia. They then burn off the offending regions. (The tissue in the infarct scar, which is already dead, does *not* itself cause the arrhythmia – it is the surrounding damaged tissue that causes the trouble.)

Trayanova's group took MRI images of infarcted pigs pre- and post-ablation and used the data to set up a unique model of each pig's heart. In the experiment, some of the ablation procedures had been successful at preventing arrhythmia and others had not. Trayanova's model identifies which regions of the pig heart are causing the arrhythmia and determines whether the ablation targeted the correct region. In five pigs, the model 'predicted' (retrospectively) which ablations had been successful and which had failed. In one case where a pig had multiple ablation sites, the computer identified which one had actually done the job and which ones were unnecessary.

Trayanova's team is now conducting a similar retrospective analysis of human ablation procedures. One of the challenges in human studies is that the clinical MR images are of

A. Bruaset, A. Tveito (Eds.), *Conversations about Challenges in Computing*,
DOI 10.1007/978-3-319-00209-5_7, © Springer International Publishing Switzerland 2013

lower resolution. If the retrospective human studies are successful, Trayanova hopes to use the computer models to guide ablation *before* the ablation is actually done. Prospective animal studies will need to first be performed before attempting any human studies. She says that if the methodology is indeed successful, it may take some time for this procedure to be approved by the US Food and Drug Administration, because the model would be considered a Class III medical device (no previous comparable technology). 'My dream would be to see it used clinically in three years', she says.

In 2011, Trayanova's lab collaborated with a team led by Dr. Colleen Clancy of the University of California Davis on developing a computational framework for a virtual drug screening system where drug–channel interactions are simulated and the effects of drugs on emergent electrical activity in the heart predicted. The proof of principle was their study published in *Science Translational Medicine*, where the computer model was used to compare the pro-arrhythmia potential of two anti-arrhythmic drugs, lidocaine and flecainide. Both drugs work by blocking sodium channels. 'In a single cell, both drugs did not appear to have any overt proarrhythmic potential', Trayanova said. However, clinical studies elsewhere had shown that flecainide actually *increased* arrhythmia in patients who had pre-existing structural damage to their hearts. The risk of sudden cardiac death for patients taking flecainide increased by a factor of two to three, compared to patients taking a placebo.

Trayanova's computer model explained why. In a simulated heart with lidocaine, the heart returned to its normal rhythm after a premature ventricular contraction. But in a simulated heart with flecainide, premature ventricular contractions tended to push the cardiac cells into an abnormal rhythm, somewhat like people awakened from their sleep by an alarm that goes off at the wrong time. In effect, each cell would tell its neighbour to wake up and the result would be a wave of unsynchronized electrical activity propagating around the heart instead of a normal, synchronized heartbeat. This understanding would not have been possible without a cardiac model that integrates events at the cellular scale with tissue-level and organ-level phenomena.

Trayanova began developing her computer models at Tulane University, but she comments that the cardiologists at the Tulane hospital were 'not very receptive to computer modelling' (she was in the school of engineering at the time). In the wake of Hurricane Katrina, several other universities called her to ask if she wanted to relocate and rebuild her laboratory. She was especially impressed by the eagerness of the doctors at Johns Hopkins to put *in silico* models to clinical use. 'I wanted to do clinically relevant work', she says. 'That is the reason I moved from Tulane to Hopkins'.

The following interview was conducted on December 29, 2011.

▶    **How did you get started in science? Did your education in Bulgaria give you a good start?**

I came from a very academic family. My father was a professor and director of the Institute for Biophysics in the Bulgarian Academy of Science. My mother is an economics professor. It's one of these cases where I didn't know of any other career path. My sister is also a professor in the Department of Material Science at Johns Hopkins. She has a different name, so not many people know we are related. The science education in Bulgaria was actually very rigorous.

▶    **When did you leave Bulgaria for the United States?**

I came to the United States for the first time at the end of December 1986, on a fellowship from the National Academy of Sciences for students from Eastern Europe. It was very serendipitous in a way, because my dad was a visiting scientist in the US in 1964 at MIT. He wrote several books and I have met a lot of people in the US who knew him. He worked in the field of motor control and a group in Arizona worked on similar problems. Some of them came to work in his lab in Sofia, Bulgaria, in the mid 80s. From casual contact with these people, I learned that there were fellowships you can apply for to study in the US. There was zero information available at the time, so it was only through the US researchers that I could find out how to apply.

I came to Duke in January 1987 and I stayed for a year and six months, went briefly back to Bulgaria, then came back again to a longer-term position at Duke.

▶    So the second time it was your intention to stay permanently?

It was all funding dependent. I was a postdoc and I wrote my own grant, a Whitaker Foundation grant, which got funded. I started teaching in the department of Biomedical Engineering. Duke was really important for my career. I worked initially with Dr. Robert Plonsey, who was the person I decided I wanted to work with when I was in Bulgaria and got the fellowship.

▶    What did he work on?

He was the first to develop a formal approach to bioelectricity, electric fields in biological tissues. His research was theoretical, on models such a single fibre in an infinite medium, but it provided a strong underpinning for the modelling that we do now.

▶    When did you leave Duke?

I accepted the position at Tulane and started on January 1, 1995. I went from a research assistant professor at Duke directly to a tenure-track associate professor at Tulane. I brought my Whitaker grant with me. Tulane was the first place where I had a large lab.

▶    Is that where you started cardiac modelling?

No, I started doing cardiac bidomain simulations[1] at Duke with Plonsey. People who worked with Plonsey did either cardiac or neuroscience research, because the approaches are similar in one dimension. I knew I liked the cardiac stuff when I was at Duke and I linked up with an experimental cardiac group that was led by Raymond Ideker.

▶    How much mathematical background did you need?

I have an undergraduate degree in theoretical physics, so I did a lot of math in my student years. When I came to Duke, I worked mostly on analytical solutions to these problems. Since a single fibre is a cylindrical object, I used Bessel functions and classical math. I was pretty good at math at that time.

For my Ph.D. research in Bulgaria I studied skeletal muscle fibre biopotentials; thus it was not hard to switch to research on cardiac bioelectricity. However, I took zero biology in college – I am self-taught. I have no formal education either in biology or in physiology. I like to joke that there are two ways to learn a subject: Either you teach a class on it or you write a grant. I started teaching a bioelectricity class at Duke and that is how I expanded my knowledge of biology and physiology.

In the US, I found out immediately that my success depended on how well I knew biology and physiology. No matter how good your mathematical model is, you won't get anywhere if people don't understand your model or buy into what you're doing. It took me some time. I had no mentor, I didn't have a PhD advisor to guide me. When I teach students now, I give them advice on every aspect of their life. I make them write grants and teach them to write papers. I wish someone had told me how to write grants or how to pick a problem that's important.

I think that not having an academic pedigree was, for me, an initial barrier to advancement, but in a way it turned out to be an advantage because it made me so much more of a keen observer. I learn from my failures quite well. I get depressed but I don't get discouraged.

▶    Can you give me any examples of failures you learned from?

---

1   Bidomain models simulate the intracellular and extracellular media as two separate domains that overlap in space. They have become standard in cardiac modelling.

You can start with my first NIH grants. I had a Wall of Shame in my lab. We would put everything that was rejected on the Wall of Shame.

To some extent the level of rejection comes with the field. Computer modelling of the heart is a field where it is so easy to fall into the cracks, because you're neither a mathematician nor a cardiac physiologist. It often happens that a mathematical model or a simulation is not noticed or appreciated by the people who do experimental work on the subject.

Another place where I was always rejected was cardiology conferences. I figured that in order to get my simulations noticed, it was important for me to go to cardiology conferences, not to mathematics or engineering conferences. The one conference I attempted to get into in the early days is now called Heart Rhythm Scientific Sessions, but previously the organization was called NASPE, the North American Society for Pacing and Electrophysiology. We always sent our abstracts there and they would always get rejected and end up on the Wall of Shame. They would get rejected because it is predominantly a society of clinicians. If you don't write the abstract in such a way that a clinician or an experimentalist can understand it, of course you're doomed. That was the lesson I learned from these experiences.

▶    How many grants did you have on this Wall of Shame?

Probably not that many; we had many more abstracts than grants. In my first year at Tulane, I had all my grant applications rejected. By the second year I started figuring out how I could write them better and things started to pick up. With abstracts it took a while longer until we found out how to express our findings in terms that experimentalists and clinicians could understand.

▶    In Oslo you said that you were very excited to get two papers in Science Translational Medicine in 2011. Was that another thing you have been trying to do for a while?

My excitement had more to do with the general idea of who I am, who I want to be, and where I want my research to go. One of the reasons I wanted to leave Tulane, Katrina or not, was my lack of clinical collaboration. I really did not like that. At Hopkins I immediately got involved in research with electrophysiologists.

When people talk about translational medicine, the term is understood to mostly mean that you look for small molecules that are targets for drugs. To have a simulation approach considered something that could be translated to clinical practice is, to me, very exciting as a concept.

▶    What was the main work you did at Tulane?

When I was at Tulane, the majority of my research focused on defibrillation. At the time, there were several other groups doing simulations on a similar (tissue) level, but they didn't move much from that level. Somehow our team got ahead. The reason is that these are complex simulation problems. You have to do great science, but you can never forget tool building. I love the science part and I want my papers read by clinicians and physiologists, but I had to keep the pedal down on the tool building and the methods. The only way that you can come out above everybody is to keep working on the methods as well as the applications.

▶    Could you talk about the tools? What distinguishes your models from the other computer models?

NT: My lab was the first to develop a whole heart bidomain model that allows one to examine how defibrillation works in the organ. There are two equations that describe the current flow through cardiac tissue. One is a reaction diffusion equation and the other is an elliptic equation. The inclusion of the elliptic equation makes the problem much harder to solve. In recent years, my lab teamed with Drs. Vigmond (my first postdoctoral fellow, recently relocated to the University of Bordeaux) and Plank (a visiting scientist from the University of Graz, Austria, in my lab from 2006 to 2008), both of whom are experts on tool building. We currently use a bidomain solver developed independently by them. Ours is a three-way partnership that we all cherish very much.

Also, when I moved to Hopkins, we started to work on reconstructing the heart's geometry and structure from MRI or other imaging modalities. Previously, two or three geometric models of the heart had been published, based on histological reconstructions, and we were using these geometries. I thought this was very constraining and I really wanted something that came straight from MRI but I didn't have the technology to do it. So I hired an image processing scientist to develop the pipeline for heart model construction and now it works. Otherwise I couldn't see how we could do an individualized approach to cardiac modelling, which is my vision.

▶  In Oslo you mentioned this idea of a Google heart. What are the advantages of having the doctor be able to see a model of a patient's heart?

One of the roles of computer models, I believe, is that they are an infrastructure that can hold information. You can include geometric modelling of the heart, equations at the cellular level as well as of subcellular processes, and regional differences in properties. All of that can be held within a model. Information about genetic disorders in ionic channels can be incorporated into Markov models.

There aren't currently equations for all of the processes. The hope is that as we dig down and obtain new information, this can be put into a formal mathematical expression. You can decide to change a particular gene or protein behaviour and use the model to see the emergent behaviour at the level of the organ. Or you can change several properties and study how their interaction manifests in the function of the whole heart. You can alter the processes at any level and see how function at the other scales of biological organization changes.

▶  This seems like a very 21-century view of medicine. If you have new discoveries, you can hang them at different points in this model.

That's what I would like to see happen if I am alive to witness it. There is no other tool, at least known to us now, which can hold information in a hierarchical structure, incorporate interactions, and predict behaviour.

▶  You talked in your lecture about needing to get the time to construct a model and execute the simulations from one to two days to one to two hours. What are the strategies for getting information out that much more quickly?

For the next 10 years to come, probably it will be dependent on the specific clinical application. For ablation of ventricular tachycardia, the application that I talked about, the way we are going to speed up the modelling is to determine how much of the cellular information is needed. One of the reasons we did this project first was that some of the electrophysiological information at the cellular level does not need to be known in detail. For instance, we may not need to calculate the detailed calcium cycling [within the cell].

That may not be the case for other applications. For the drug study, we ran the models for a long time. One simulation would run for a week. These are very complex Markov models of drug–channel interactions. The drug can be charged or non-charged and it interacts differently with the sodium channels. We expect to have an extensive library and have been working on building one already, of a variety of cellular behaviour models that one can choose from, a library that runs the gamut from very simple to very complex. Also, methodologies to automatically segment patient hearts from MRI scans would provide significant speedup for patient-specific applications.

Another simulation speedup strategy is to run our models on graphics processing units (GPUs).

▶    Why are GPUs better?

One advantage is that they are very small. Our computational resource (cluster) is in a big room at Hopkins and it requires a huge amount of energy to be air-conditioned. If all simulations could be run on a desktop machine, it would require no storage space, no specific climate control – and you could put it in a doctor's office!

Another advantage of the GPU is the much higher floating-point performance (three and 10 times higher than CPUs for double and single precision) and the higher memory bandwidth (by a factor of 10). Both flops as well as memory bandwidth are key for scientific applications such as the simulation of a heartbeat. Therefore GPU execution could be roughly 10 times faster than CPU execution if one manages to implement the code properly on GPUs, which could be tricky. In the lab of my collaborator, Dr. Plank, we have already a prototype of a full-blown GPU-enabled heart simulator that delivers excellent performance on small GPU clusters. The initial results suggest that we will have the performance of a mid-scale supercomputer integrated in a desktop. This is what is needed to leverage modelling in clinical applications, by driving down the costs typically resulting from traditional supercomputing approaches.

▶    Since your team studied the proarrhythmia propensity of the drugs lidocaine and flecainide, have other people asked you about other drugs they want to test? Will this move from a proof of principle, that modelling can predict unexpected effects, to a standard practice?

This is difficult to answer because pharmaceutical research is a big industry. In order to make a good prediction of what a drug does, one needs a good model of its interaction with a particular protein or channel and a way to examine the properties resulting from this interaction. Dr. Colleen Clancy, who is the last author on the drug safety *Science Translational Medicine* paper, has always been interested in developing Markov models of ion channel behaviour. It was a really great opportunity to combine the drug interaction model they had developed with our human heart model to demonstrate the emergent properties at the organ level.

In general, we are increasingly getting involved in research that I never would have thought I would be involved in. Science has become so collaborative. We all search for what is the best way to address a particular problem. Dr. Clancy and I are now applying, together with Dr. Robert Harvey from the University of Reno, Nevada, for a grant to study how the distribution of nerve endings in the heart affects the propensity to arrhythmia. I never would have thought of doing that myself.

▶    Statistics has become such a widely accepted tool in medicine. Could mathematical simulation become equally accepted, at least in some areas?

That is my personal mission. I really care about that very much. I have learned how to present my case to doctors so they don't feel this is above their heads. I have made a major effort to learn the clinical side of the story.

▶    Let me ask you about the other 2011 paper from Science Translational Medicine, about using a high-frequency alternating current for defibrillation. Could it really improve on the defibrillation methods that are used today?

I've always been interested in defibrillation. It's an area that I know very intimately, because I spent many years researching it, but I found that when I worked in this area it was really hard to make a difference, to make people listen to me.

I love teasing out relationships, and there is a lot of intellectual pleasure in doing that. But then some time passes and you say, so what? Where do we go with that? If my studies are only to discover how the shock interacts with the cells and nothing ever changes as a result, then I might get papers published, I might get grant money and get promoted, but still, to me, intellectually, it's not really satisfying. And what I found with defibrillation was that we learned a lot, but it was very hard to make a change in the way defibrillators work nowadays.

That's why I decided to drastically decrease the involvement of my lab in defibrillation research. I went from a lab in which we had 15 people working in that area to a lab where I now have two. In the paper you asked about, we attempted to do something radically different. My co-authors, Drs. Berger and Tandri, had this idea of using high-frequency electric fields to cause conduction block in the heart. This [alternating current] idea presented itself as an opportunity to do something new. I had already written two papers on the response of cardiac tissue to alternating current in two-dimensional models, for various frequencies, so this naturally became a very interesting collaborative project. I still have a student working on the project.

▶    Where is the project going?

We sent a pre-proposal to the Coulter Foundation and we got invited to submit a full proposal. We also have a patent on it. It's still not a low-voltage defibrillation; it uses high voltage, but it's much better than normal defibrillators in terms of the post-shock dysfunction. It also offers the prospect of a pain-free defibrillation since the high-frequency field used to defibrillate the heart can also inhibit nerve conduction. Again, we have to test it in large animals. The project is on-going.

▶    You have also started a company recently, CardioSolv. Can you tell me what the company is working on?

We started about three years ago. We have a group of five co-founders, all of whom came through my lab in one way or another. The three faculty co-founders are Drs. Vigmond and Plank and myself; I talked about my partnership with Drs. Vigmond and Plank earlier in this interview. The company webpage, cardiosolv.com, provides information about our roles. Dr. Vigmond is the president, Dr. Plank is the chief technical officer, and I am the chief scientific officer. The only current employee, Dr. Brock Tice, is a former student of mine.

At this point we are selling software and providing services. We have sold licenses for our software to several universities and have had industrial contracts. The long-term goal is to bring personalized modelling of the heart to the market. That was the whole point of starting the company.

▶    What made you think that this was the time to start a company?

We were able to do individual hearts and I also had a very entrepreneurial student, Brock. Somebody has to push you. He was very motivated to do it. He and I were talking about starting a company and then Drs. Plank and Vigmond joined.

▶    Simula has a strong emphasis on technology transfer. Have you learned anything from them that you could apply to your company?

Not really, unfortunately. The reason is that my interactions with Simula have been primarily scientific. Now that my former student Molly Maleckar works there, my relationship with Simula has strengthened and I have a much better vision of what they do. We have a very different technology transfer process here and I would love to learn from their experiences.

# As Simple as Possible, but Not Simpler

*An Interview with Alfio Quarteroni by Dana Mackenzie*

For at least two decades, Alfio Quarteroni has been at the forefront of research on fluid dynamics, especially in the most complicated type of problems: those in which a moving fluid interacts with a flexible surface. He is perhaps most noted for his role as the chief scientific advisor for the Swiss team that won the America's Cup yacht race in 2003 and defended its title successfully in 2007. The team's secret weapon was computational fluid dynamics, which they used to obtain the largest possible performance improvements from tiny tweaks in the shape of the boat.

Quarteroni has also studied blood flow in the human circulatory system for several years. Like the water flowing past a yacht, the blood interacts with a deformable surface, in this case the artery walls. The problem in this case is made even more complicated by the vast range of scales in the circulatory system. Very large blood vessels such as the aorta must be simulated by three-dimensional methods based on the Navier–Stokes equations. Smaller vessels are too numerous to be simulated in full three-dimensional detail, but they can be approximated as cylindrical tubes in which the blood flow is described by one-dimensional equations. Finally, the billions of capillaries are too numerous even for this one-dimensional approach. Instead, they are lumped together as one entity with properties such as resistance and inductance, which are modelled on electric circuits. These are called zero-dimensional or 'lumped parameter' models. Linking together these three very different types of models remains a daunting mathematical and computational task. 'It takes about eight hours on a supercomputer to simulate a single heartbeat that lasts one second', Quarteroni said in his lecture at the Challenges in Computing conference. 'We need to invest in more efficient ideas so that doctors can use these simulations'.

One promising class of techniques that Quarteroni discussed is called model reduction. As he explained, the idea of these methods is to 'simplify the partial differential equation (PDE) without compromising the fidelity of the model'. One example he described was a semi-empirical approach called surface registration. Instead of simulating both the blood flow and the arterial wall's motion from scratch, you can obtain observational data from an MRI scan and use that as a prescribed model for the wall's motion. In effect, you let nature solve half of the problem for you and then use computation to solve the other half.

Another model reduction technique, discussed in the interview below, is to solve the computational fluid dynamics problem only for a select group of model patients and then describe every new patient as a sort of weighted average of the model patients. Then the modelling of the new patient involves solving differential equations of much smaller size. The goal (not quite achieved yet) would be to make this procedure fast enough for a doctor to do online in his office.

Quarteroni has used this and other model reduction techniques to design coronary bypasses and ventricular assistive devices. 'It takes us perhaps 20 hours, when it would take

A. Bruaset, A. Tveito (Eds.), *Conversations about Challenges in Computing*,
DOI 10.1007/978-3-319-00209-5_8, © Springer International Publishing Switzerland 2013

700 hours with standard finite element methods', he said. However, some patients may not even be able to wait 20 hours. Improvements in model reduction will be necessary before these models can be used in clinical practice.

The following interview was conducted on January 14, 2012.

▶    How was your Christmas vacation?

I had many different letters to write. I had to assess 48 big grant proposals for the European Research Council. I myself received one of these big grants three years ago, so of course I felt obligated to do a serious and responsible job.

▶    What's the name of the grant?

It's called an Advanced Grant for Senior Researchers. There are very few but they are very substantial, up to 3 million euros. Our project is called MathCard, mathematics for the cardiovascular system. We will model the arterial system and heart with numerical simulations, validated against clinical measures. We now have 15 young researchers supported by this grant, PhD students and postdocs in Lausanne and Milano.

▶    How did you get started working on the cardiovascular system? Was it the first computational math project you worked on?

It was not my first project. I started my career as a theoretical mathematician. Then, after a few years, I had the possibility to visit NASA Langley in Virginia, where there was a very good centre of scientific computing named ICASE (Institute for Computer Applications in Science and Engineering). I got involved in applied mathematics there, including fluid dynamics. Later on, I branched out into wave propagation, specifically seismic wave propagation. Another project I worked on was modelling the process of the diffusion and propagation of pollution in the water and the air. And there were several others.

I started working on the cardiovascular system by pure chance. I had two completely independent contacts. One was a young Italian heart surgeon, Massimiliano Tuveri, who was interested in deriving simple elementary equations on the wave pressure that arises after a certain intervention in the heart. There was a basic issue on the behaviour of fluids under a pulsatile force. Of course I didn't have any reasonable answer to give him, so we started making some computations. It was the first time it came to my mind that blood flow was important and could be addressed mathematically. Meanwhile, I had a fresh PhD student in Milano who in the course of his master's project had done an internship in a hospital. In this hospital, one of the doctors was interested in computing the geometry of the carotid arteries. After this student started at Milano, he took a course from me, not in blood flow but in regular PDE, and he asked me to be his advisor and said he would like to work in blood flow modelling. His name is Alessandro Veneziani and he is now a professor at Emory University and doing a superb job with medical doctors.

I tried to connect the two inputs that I had, from the student and from the surgeon who was living a thousand kilometres away. This was roughly in the mid-1990s. Of course, a whole set of mathematical challenges and difficulties opened up. The very first thing you try to do is use the Navier–Stokes equations on vessels with rigid walls. Then you are tempted to see what happens when you have some malformation because of plaque deposits, narrowing because of atherosclerosis. Then you want to see what's happening when you include the compliance of walls. On one hand, you have to develop some understanding of the differential equations on the theoretical level and, on the other hand, you have to set up numerical algorithms to capture what's going on.

▶    Obviously you weren't the first person to try these things, because you mentioned lumped parameter models and one-dimensional models. What parts of the previous work did you find most useful and in what ways did you have to go beyond what had been done before?

What we discovered in the literature was a huge amount of low-dimensional models. Biomedical engineers had done lumped parameter models since the 1970s. That's a good way to capture the fundamental behaviour of blood flow in the circulatory system. You don't care about the details, such as the way blood interacts with the vessel wall or the way it may become unstable past bifurcations, but you have a global picture. It's a coarse picture but it's a visible picture. One of the basic problems was how to connect these lumped parameter models, this background description, with a more sophisticated description based on three-dimensional analysis, which is more appropriate for the fine scale.

This is already a challenging mathematical problem, because you are coupling two different kinds of descriptors or operators. Apparently this was not done at that time. This was new. Besides, there was this other intermediate way to describe the situation through one-dimensional models. Indeed, Euler himself was the first person to propose a one-dimensional model to describe blood flow in cylindrical vessels. I think we were the first group to find a systematic way to set up this coupling and to put it on a rigorous and systematic and solid mathematical foundation.

Of course, today we are still using this approach in a much more sophisticated way. We can try to incorporate real data from real patients in the model. It's one thing to set up a mathematical paradigm or formalism, but another to try to use it efficiently in a real, physical context. For instance, you have to account for the external tissue. These arteries are not living in empty space. They are in our body and what is around the vessels is important because it's interacting with the dynamics, the displacement and deformation of the vessels. There is another world of complexity you have to account for. One very essential point here is how to simplify this complexity without compromising the reliability of your computation.

▶   It struck me that the lumped parameter model is one way to reduce complexity, the old-fashioned way. Are the new ways better? If so, how?

I would not say it is an old-fashioned way. It is in the sense that it has been there for 30 years. But in mathematics you can have wonderful ideas that last for centuries.

The lumped parameter model is a geometrical reduction. In our body we have tens of big, large arteries and tens of thousands of small arteries and millions or billions of capillaries. There is no way to overcome this forest of little vessels and capillaries without using a very simplified approach. We call it a geometrical multiscale approach. On the other hand, this may not be enough to face some problems. The complexity reduction strategy you should use is very strongly driven by your goals, by the problem that you really want to solve.

To give you an idea, if you want to solve the problem not only for a single patient but for a class of potential patients, you want to consider healthy people with different sizes or weights or ages. The basic mechanisms behind blood flow modelling are the same because you have the same equations, but the geometry has different parameters that describe the stiffness and elasticity of the vessels and the viscosity of the blood. They vary in a continuous way, not a discrete way. So how can you account for this variation without solving one new problem for every specific choice of parameters?

This calls for a different kind of reduction. You have in the background a geometric multiscale model, but you have so many possible parameters that you would have to carry out thousands of simulations. If it takes a couple weeks to simulate just one physical second of a single heartbeat on a supercomputer and if you have thousands of patients, it will take you 2000 weeks. That's ridiculous! We can tackle this problem through model reduction. In my lecture I talked about the reduced basis method, which works something like this.

Let's consider again this set of parameters and assume that you take 20 or 30 representative combinations of those parameters. You assume that these 20 or 30 combinations somehow characterize individuals who belong to a specific category, for example, a young adult male. For each of these 20 or 30 combinations, you carry out your computation that lasts several weeks on a computer. Now, what happens if you have a class of thousands of individuals? You have individual number 728 with substantially different characteristics but you'd like to see if you can use the previously computed solutions to characterize that individual. We call these previously computed solutions snapshots. You project this new individual into the subspace generated by the snapshots.

What is the advantage? What you are basically doing is separating your approach into two steps. One is offline and the other one is online. The offline step is what you do for the 20 or 30 representative individuals that you decide to consider in your database. For every one you have to use your blood flow models, based on 10 million or 100 million unknowns. That's a big cost but it's done once and for all offline. Then you generate your subspace with these precomputed snapshots and with each new individual you just have to project to your subspace. This is a low-cost approach that is carried out online. It might take a few seconds or a few minutes. You can carry out investigations for many individuals based on a reasonable number of individuals that are sufficiently representative of the family you are considering.

This approach works not only in blood flow problems, but also in aeronautical engineering to design a new airfoil. You have many parameters and you want to explore the complete set of parameters by taking advantage of solutions that you have precomputed.

▶      If you're using these spaces of previously computed solutions, are you assuming linearity? Is that a problem when you have very nonlinear equations?

That's a very good question. The answer is yes and no. You assume linearity in the sense that you project into a linear space, but in fact the problem at hand can be nonlinear, which is actually the case for blood flow.

▶      How well do these linear combinations match what you would get if you worked through the individual solution?

That's also a very good question. In fact, the reliability of our computation depends strongly on the way we choose the individuals in the beginning. If you want to get information on the way cardiovascular disease develops in the US and you take 20 individuals that are from different countries, from different continents, they may not be representative at all. You need to take snapshots that are sufficiently representative and you need to base these on precise and rigorous procedures.

We use an adaptive choice procedure. You want to exploit the fact that your mathematical apparatus can predict the kind of error you are making. Therefore, if you fix *a priori* the tolerance that you want to allow, then your algorithm suggests the way to select the $(n + 1)$th individual once you have selected the first $n$. It is an adaptive procedure based on control of the error that you are generating. This type of paradigm or philosophy is very common these days in numerical analysis. Many accurate methods are based on *a posteriori* error estimates.

▶      Have you used this same approach in other problems? It seems to me that in the America's Cup problem you could vary the shape of the hull or the keel.

Not really, for two reasons. The first point is that when we worked last time on America's Cup, in 2010, we were not advanced enough with this technique. The second reason why we did not use it in that specific context was that in America's Cup the flow of the water around the boat's appendages and also the air around the sails are turbulent. In ideal conditions in our body, the blood flow is not turbulent. When the flow becomes turbulent, the problem is more difficult. If I may oversimplify, you depart very substantially from that implicit assumption of linearity that I was making before, so the result may not be very accurate.

▶      So the problem does have to be fairly well behaved in some sense?

The underlying assumption, the moral assumption you might say, is that if you vary the parameters, your solution should depend smoothly enough with respect to variation of those parameters. In problems where this is the case, the reduced basis approach can work well.

▶     In Oslo you mentioned two applications of your cardiovascular work, coronary bypass surgery and ventricular assisted devices. Can you tell me more about these applications?

In both cases it is a matter of creating a prosthetic implant. When you apply a prosthesis, you are changing the blood flow quite dramatically. For instance, in the case of the coronary bypass, you have a coronary artery, occluded totally or partially. You want to create an artificial external bridge. You want to insert a vessel that is sucking blood upstream from the occlusion and bringing blood into the main branch of the coronary artery downstream from the occlusion.

Downstream, at the reattachment point, you may create a vortex. The vortex depends on the geometric features of the bypass: for instance, its diameter, the length, and the angle of reattachment. Ideally you would like to minimize the vorticity, because it might increase the shear stress, which is the force the blood exerts on the wall. If the blood contains too much cholesterol, say, this might cause deposits on the wall over the long term. After a few months, you might have another potential occlusion downstream because of this recirculating flow that has been artificially induced by the bypass.

Minimizing the vorticity is a problem of optimal shape design, like drawing an optimal airfoil that maximizes lift and minimizes drag. You can use the same mathematical approach to designing the ideal shape of the bypass. There are many parameters that describe the bypass. Again you have a problem with many parameters and you might want to use this reduced model approach to reduce the complexity. In this case, if you want to optimize something, you have something that mathematicians call control and optimization, in which you reach the final solution by an iterative process. At every step you have to solve a very complex problem with many parameters. This is where the reduced model is coming into play and is extremely useful.

The ventricular assisted device is a similar story. A child has a ventricular deficiency – for instance, the blood is not pumped at the right flow rate outside the heart – and you need a kind of artificial external pump. You have to inject blood from this pump into the ascending aorta. You will create a major perturbation in the ascending aorta, which is the main conveyor of blood. Again, you want to optimize the shape of this ventricular assisted device. You have to control the geometric parameters and you also have to control the flow rate of the external pump, so you have other parameters that enter as boundary conditions in your system.

▶     How close is this to being useful in a clinical application?

It's still very far if you think of systematic use. It's far from becoming protocol. On the other hand, we have several direct contacts with cardiovascular surgeons. They do this surgery on the grounds of their own knowledge and experience and intuition. What we are providing them is perhaps a more quantitative approach. It's a way to provide them with a complementary viewpoint with further information that is quantitative: If you change this angle by 1 percent, then the variation of pressure in your coronary artery might be 3 percent or 10 percent. Those things are extremely difficult to guess without mathematical support.

▶     So it's not at the level of individual patients yet, but more at the level of broad design and understanding how these devices work?

This is correct, although the more you know about the individual patient, the more reliable the model will be. It's not intrinsically impossible to go to the level of the individual patient just because you might be missing some data. As soon as you have the data, the model is capable of reacting.

Of course you should never forget that these are living tissues and living organs. It's not like the flow around an airfoil or an America's Cup boat, where, in principle, the design is completely known and it's only a matter of how many resources you want to devote to this problem. If you have a sufficiently good computer in your hands, you know you can satisfy very severe accuracy requirements.

In the cardiovascular system the human variability is so big and there are uncertainties. After all, the patients do not react as machines. Two patients who appear similar will react in completely different ways. Still, you are providing the right indicators and these indicators can really help the surgeons to devise their own strategies of intervention.

> ▶ How good were these devices before you looked at them? Presumably they were designed by hand and by intuition.

They are very close to optimal. When doctors do this, they end up with shapes where the 'optimal form' is there qualitatively. On the other hand, don't forget that in these problems, the solution is extremely sensitive to shape variation because of the intrinsic sensitivity of the flow field on the boundary. Two shapes that look similar to a specialist in the field may behave in very different ways as far as the interaction with the flow field.

This means that when you look at what clinicians develop on the basis of their own experience, although it resembles closely what a mathematical model would indicate, nevertheless tiny differences might have a very important role. This is the same experience we had with the America's Cup. The winning team does not have a revolutionary design – not any more at least. It is able to improve by making tiny increments, tiny changes that might have a very important consequence.

> ▶ One of the other speakers in Oslo, Natalia Trayanova, also discussed the cardiovascular system from the point of view of modelling the electrical fields in the heart. Is there any possibility of putting her models together with yours?

So far, we have been talking about the way the blood behaves, but the main engine is the heart. How do you model the heart? This is an extremely challenging problem indeed. The heart, if you oversimplify the story, poses a three-field mathematical problem. One field is fluids, because you have blood in the atria and ventricles. The second is solid mechanics, to describe the myocardial tissue that deforms. Finally you have electrophysiology, for the electrical waves propagating in the heart. The electrical waves trigger the deformation of the solid tissue, which interacts in a nonlinear way with the blood. All three fields are very strongly interconnected.

In the long run, of course, you would love to have at your disposal a coupled model that is capable of combining these fields. At present, what you often see is an approach that is extremely advanced in one of the spheres, such as Natalia's. She was primarily talking about the electrophysiology of the heart. Of course you can study this independently of the two other fields. You can assume that the tissue is steady and there is no blood, that you are only simulating the chemical and electrical components of the story.

On the other side there are people such as Charlie Peskin from Courant, who started in the 1960s with a very new and original method to simulate blood flow in the heart. His great and seminal idea was to consider the muscle, the tissue, as part of the flow region. His approach, the immersed boundary method, accounts for the other two components, the fluid and the mechanics problem. Both of these approaches [Trayanova's and Peskin's] are very advanced but can be regarded at this stage as, to some extent, independent. The challenge is to couple them together, which is coupling three fields together.

This is where you have a major challenge, because it is not just a matter of finding the right mathematical model. The first challenge is to address the physiology. The second challenge is to address at a numerical level how you solve this problem on the computer. There are different time scales and length scales. It's especially difficult when you consider pathological events such as arrhythmias or fibrillations.

I would say there are a few groups trying to address this challenging problem, but we have not yet arrived at a satisfactory answer for the three-field problem. However, we do already see some preliminary solutions to this problem. Again, it's a matter of establishing your goals. If you want to use this model to feed the circulatory system, then we are almost there because you don't need to see the fine details of the heart itself. You want to see the way this will eventually produce a blood flow rate and a pressure pulse in the descending aorta. The output is simple compared to the complexity of the whole machine. We're just measuring one effect of the whole machine. I think we are not far from solving the three-field problem for this purpose.

On the other hand, if you want to use the three field problem to understand heart pathologies, where the heart valves behave in an imperfect way, or you want to predict a stroke, then, in that case, you need to rely on a very accurate model. I think we are still relatively far away from that.

▶  That leads very well into my last question. Looking at the challenges ahead, what problems do you think should be solvable within 10 years and what problems still need new ideas?

If we talk about life sciences, say, certainly one deep problem is the one of combining different length scales, going from the cellular level up to the tissue and the organs. Until now I have been describing what is happening at the organ level, the systemic level. Then you have to go down to the mesoscale and the microscale. If you want to see how atherosclerosis develops, you need to go to the level of cellular behaviour, the endothelial cells, and study how they are damaged by blood forces. Of course, you want to create a connection between scales.

The multiscale problem is present not only in this context but also in other sciences. There have been an incredible number of initiatives in the academic environment worldwide to face multiscale problems. I believe that in 10 years we will see major, major advancements in this respect. The coupling between the scales is definitely possible in a 10-year time frame.

Of course there are other problems that might be more demanding. For instance, people are now working on the functional genome and the epigenome. The number of degrees of freedom required to understand those things is so large. This is almost unexplored territory for mathematics, I would say. There is room for brand new ideas. We are speculating on new laws of behaviour, which will not just be deterministic laws; they will be stochastic. We need a gigantic effort to start shedding light on this problem.

▶  It sounds to me like trying to model the cardiovascular system if you didn't know the Navier–Stokes equations.

Exactly. When you are simulating blood flow in the human circulatory system or aerodynamics around an airfoil or seismic waves in the earth, so many simplifications have been made in order to obtain the equations: Navier–Stokes for water flow around a boat or the elastodynamic equations for waves propagating through the ground. Yet you end up with solutions that are very, very close to what you measure in practice. That is really an incredible fact. In spite of the fact that you oversimplified reality, those equations are bringing so much information and they are capable of reproducing the essential features of the phenomena you are interested in.

That is what you miss in biology, and what you miss when you are facing problems such as those I was mentioning before. You need to find which fundamental principles are governing the tremendous complexity that you observe. You don't know yet the best way of simplifying reality. This is where substantial and fundamental achievements remain to be made.

# A Caring Critic

## An Interview with Magne Jørgensen by Kathrine Aspaas

'Pity the leader caught between unloving critics and uncritical lovers,'
John W. Gardner, US Secretary of Health, Education and Welfare and President of the Carnegie Corporation (1912 – 2012)

We are surrounded by students in the university library cafe. Have elbowed our way through the crush in the middle of the exam season, with the lilacs nodding their rain-drenched purple heads through the big picture windows. The students are unaware that when – in the not too distant future – they join the world of work, the way they work will have been altered by the insights developed by the tall, amiable man drinking coffee beside them. They are all major consumers of computer systems, and – who knows – maybe some of them will one day be responsible for the procurement or design of really big computer systems.

If so, they will need to know that they are far from being as logical and strictly analytical as they would like to think. They must recognise that the thirst to win a contract or acquire equipment at a knock-down price often gets in the way of unbiased analysis.

The core hypothesis in Magne Jørgensen's research is that we have a naive view of our own rationality, which goes a long way to explaining why major IT projects take on average 30 to 40 per cent longer than estimated to develop. What lies at the root of this over-optimism?

Let's start with the anchor – the so-called anchor effect. Revealed by the Israeli-American psychologist Daniel Kahneman, who, along with his research partner Amos Tversky, subsequently won the Nobel Memorial Prize in Economic Sciences. Their simple, but brilliant, experiments started with a number of students who were divided into two groups. A wheel of fortune was spun, stopping at a random number between zero and 100. The students were then asked whether the percentage of African countries in the UN was higher or lower than the number on the wheel of fortune. If the wheel landed on a low number, the students overall had a tendency to give a lower estimate of the percentage than if the wheel landed on a high number. Conclusion: We are influenced by the first information we receive, even when it is completely random. We believe we have a good foundation for our opinions, but they are often the result of both incorrect and arbitrary information.

## Vikings and Choral Music

For Magne Jørgensen it all started with economics and computer science at a university in Germany. At that time there was not a lot of psychology in the air, but rather a fundamental interest in the more organisational and financial aspects of computer science. However, it helps to be an omnivorous reader of non-fiction – everything from philosophy and art history to the Viking age and belief systems. This latter is possibly rooted in the Christian youth choir he belonged to at secondary school, or the Christian upbringing he received from his parents. For belief systems encompass so much more than religion, which he gave up at the age of 18. Now he is mostly interested in the psychology of religion.

A. Bruaset, A. Tveito (Eds.), *Conversations about Challenges in Computing*,
DOI 10.1007/978-3-319-00209-5_9, © Springer International Publishing Switzerland 2013

► What do you believe in?

"I am a probabilist – someone who thinks about probabilities and believes that a lot of things are possible. I reject very little, but at the same time I'm an advocate of evidence-based thinking. Lately, for example, I've read a lot about evidence-based happiness. What is it that characterises the journey from suffering to happiness? When are people happy and when are we not happy?"

► And when are we happy?

"A lot of research shows that it's actually related to helping other people and working together. This is well documented in the research, and I believe my involvement in Nepal reflects this in part. I notice that personal engagement generates happiness. I teach and mentor at Katmandu University, at the same time as we provide financial support to gifted students. Some of the money is ours and some comes from the IT industry. This kind of positive psychology certainly has its place. It can make a major contribution to people's sense of wellbeing, and I hope that this insight becomes more widely known. The research also shows that becoming caught up in something – doing something that you think is fun and that you're good at – forgetting time and place, is a powerful source of happiness."

► How did you become so interested in psychology?

"I think that understanding one's self better is part of it. It's also very relevant in project management, and the IT industry is very open when I talk about the importance of psychology when performing evaluations and making decisions. They think it is both relevant and exciting. Getting a good response from both accountants and programmers helps strengthen my interest in psychology. Take self-control, for example. When your self-control is worn down you have fewer inhibitions, and can easily make some racist remark, for example, and you are more vulnerable to external influences."

► How does self-control get worn down?

"It can happen if you have a lot of difficult conversations one after the other. If you have been arguing with your wife or children. Or when you have to sit bottling up something you want to say or do."

► Can we train ourselves to have better self-control?

"Self-control works like a muscle – without being a muscle – and can be strengthened through exposure to situations where you must exercise self-control. And postpone the reward."

► Do you do that yourself?

"No (chuckle). But it would probably be a good idea."
It is at this point in the interview that the professor gets up to buy two cups of coffee, and returns with a cake that we naturally do not eat. Neither of us goes near the treat. Not a word is said, but we are practicing our self-control.

► But this over-optimism, could it be that we are equipped with it because it allows us to believe we can achieve the impossible?

"Yes, there are good grounds for that assumption. Psychologically healthy people are in general over-optimistic. Those suffering from mild depression are most realistic, while the severely depressed can often be more pessimistic. To be a realist, it seems, you have to be a bit down in the dumps."

▶    So perhaps IT overruns are something we just have to live with?

"We do live with them, but at the same time it is worth delving deeper into the psychological processes that could keep them small. It seems, for example, that we become more realistic if we look at the past – look backwards. While we become over-optimistic when we look forwards, which we generally do when we are planning something. We frequently manipulate ourselves into thinking that it's possible to work within the customer's impossible budget. By being aware of these effects, we can moderate them. We have also developed better models for IT systems development that are more flexible and take account of the fact that projects are organic in nature, changing along the way."

▶    Are we talking about the transition from waterfall to agile models?

"Precisely. The traditional way of designing a project is first to write a specification of what you want to achieve. A wish-list of design and functionality. Then you write the program that the computer needs to perform the tasks. Finally you test whether the solution actually does what it's supposed to. Have we fulfilled the specification? This is the waterfall method, and it can work fine if the customer knows exactly what he wants at the outset, and if the solutions are clear and simple. But that's seldom how it is, because projects change along the way. That's why a better model is often to deliver a little at a time, learning from what has happened. Everyone learns along the way – the customer, the developer and the project manager. This is the agile model, which in many ways takes greater account of the project's psychology and dynamics."

## Faith, Hope and Overruns

For Magne Jørgensen it all started with the purely mathematical and analytical estimation of workloads and costs in major IT projects. Simply put, it is about large, advanced models into which you put figures, and out of which you get an estimate of the number of working hours needed. Jørgensen claims that no one has yet created good estimation models for IT development, which is why billion-dollar investments are decided on the basis of expert assessments, and much is left to faith and hope. Often all goes well, but we know that on average there is a 30–40 per cent cost overruns.

▶    You've said before that it is a good idea for project managers to listen a bit more to psychologists and a bit less to consultants with fancy mathematical calculations. That sounds as though you are actually engaged in behavioural economics?

"You could certainly place it within behavioural economics. Personally I've called it decision-making and judgment psychology, and am well within the scope of software development, which covers everything from mathematics to sociology. My last three PhD students have worked in the field of psychology. That's rather amusing because I've never actually taken a course in psychology. I study what is called over-optimism – the fact that we have a naive view of our own rationality – and there aren't many people within IT who are concerned with these psychological factors. That is why is it very easy to be a leading researcher in this area. (Chuckles)"

▶    Where are the others?

"It's got something to do with our education. We don't have many courses in psychology for those studying IT, if they exist at all. You know yourself how slowly universities move forward. There are professors there who graduated 30 or 40 years ago, when computer science was practically synonymous with mathematics. So change takes time."

▶    Yes, so we are back to behavioural research and the anchor effect. How can we learn to let go of our own ideas?

"That's a very good question. The easiest thing is obviously to carry on doing what we know – and stick to that. That's why I think it's important to be open to criticism of one's own ideas. I drill my students in this: that we mustn't only allow criticism of our work, but present our research in a way that invites criticism. That we learn to like it. This is the scientific principle in a nutshell. But it goes against human nature, so we have to work a bit harder at it. For example, I would appreciate it if you gave me some examples of my behaviour that in your view are not optimal for this interview."

▶    Fine. We'll come back to it at the end of the interview, and I hope it can go both ways.

"It's a deal."

▶    What kind of criticism have you received that has helped you move forward?

"What stung most was also what was the most fun afterwards. It was about overruns in public sector projects, and the IT magazine Computerworld's expert correspondent Peter Hidas was critical – not entirely without reason – of our research and how we communicated it. His criticism was, in my view, not wholly accurate, but I would go a step further and put in a word for the benefits of inaccurate criticism, too. It's often an indication that our message has been misunderstood, and it can tell us that we need to improve the way we present it. There is in fact very little about criticism that is not beneficial if we learn to tolerate it, or better yet, like it. Just look at me and Computerworld. One consequence of the criticism, and that I arranged a meeting with them because of it, is that I have been a columnist with the magazine for the past four or five years now. My starting point was: what is the best way to communicate our results? That's right, write about them myself."

▶    One of the things you wrote in Computerworld was that an IT tester must be destructive. Does that have something to do with it?

"Yes, we must adopt a frame of mind where the objective is to find fault. Not verify if it works. You can either find the arguments to show that it works, or look for the errors and omissions. There is research evidence that the brain operates in two modes – either confirmation mode or critical mode – and that it is very difficult to combine the two."

▶    What about sympathies and antipathies?

"Yes, it is obviously easier to disagree with someone we dislike. Being in a critical mode comes very naturally to me. I am good at finding fault. With myself and with other people."

▶    And you're happy as well?

"Yes, of course. We have two levels in our brains. In the unconscious part of the brain, which we don't notice that we're using, we can nurture the conviction that we are 'world champions'. At the same time the other, analytical part, may know that we are not. So it's possible to have an excessive faith in one's own abilities in the unconscious part, but compensate for that in the more critical, analytical part. Which brings us to the well-known internal dialogue."

▶    We're back to Daniel Kahneman and his book 'Thinking fast – and slow', which is flavour of the month in academic circles just now.

"And it's a good thing too. This is something that everyone should be aware of. I can actually see that these issues are percolating into the IT companies I talk to. The anchor effect, for example. A lot of my research builds on Kahneman's brilliant findings and methods. For example, I've studied how easy it is to form an opinion on an incorrect foundation, and keep it because you think the best arguments support it, while in reality they may be influenced by the way I've framed the question. One experiment aims to anchor the belief about whether a computer programmer has most to gain from being risk-willing or risk-averse. One of the groups argues in support of risk-averse programmers being the best, while the other half argue the opposite. I give them each an example that supports their allotted argument. Then I ask them what they think themselves, given their own experience with both risk-averse and risk-willing programmers. Those who are asked to argue in favour of risk-averseness have, as expected, a much more positive view of this quality than the others. Afterwards I tell everyone that I've framed the question in this way and given them the example in order to find out how easily influenced they are. Then I ask them to say what they think now. But it makes no difference. Their views change very little regardless. The people who cheer on the risk-averse, continue to cheer on the risk-averse – and vice versa. To see whether this lasts, I ask them again a week later. And their opinions are immovable. In their eyes they have come to their own conclusions. It doesn't help if I remind them that they have been asked to come up with a one-sided argument. The explanation is that we have a tendency to seek confirmation, and that the influence – for example from a one-sided argument – exerts its effect without our noticing."

▶    How do you manage to avoid this confirmation bias in your own research?

"Another good question… The entire scientific method is built around the notion that we must falsify instead of verify. But that's in theory. In reality most researchers are out to confirm their hypothesis and the model they have created. I try and train myself to come up with counterarguments. The scary thing is that even this often ends up with a defence of what I believe. We are by no means free of the confirmation bias. Most people are familiar with the phrase: 'we believe it when we see it'. My research shows that the reality is often the exact opposite: 'we see it when we believe it'."

▶    There must be a reason why we think like this?

"From an evolutionary point of view we probably benefit from creating patterns based on very little data. It's worse to misinterpret an existing pattern than to see one pattern too many. It's worse not to see the lion's hunting pattern – and get eaten – than it is to invent a rain dance that doesn't bring rain. We don't lose that much by dancing, even though it probably doesn't produce the desired effect. In this way we have become very good at finding patterns – in many ways *too* good. The fact is we see patterns even when faced with random occurrences."

▶    So what you are researching are human inclinations?

"Yes, and the objective is for us to become better at making the correct assessments. Evidence shows that it's very difficult to rewind the brain to where you were before you were influenced by misleading information. It's practically impossible. So the best solution is to avoid such information. In IT projects, for example, the person who knows the customer's budget should not estimate the costs. The customer's budget can easily become the anchor for the cost estimate, and that is the recipe for over-optimism. It's also a good idea to avoid loaded words when asking about an estimate, for example that the customer wants a 'minor extension'. That also affects the people who are going to frame the bid. They want so very much to win the contract. The best method therefore is to eliminate misleading and irrelevant information before the person who is going to estimate costs starts work. A lot of research has been done on so-called *debiasing*. One of my students is working on this at the moment, to find out how we can get better at removing biases in our assessments that result from misleading information."

The noise of the students around us has reached such a pitch that we actually have to shout at each other across the table. And the energy in the room is just as wonderfully intense as it is only possible to be in a student canteen during exam season. What then could be more natural than to talk about the education they receive.

## People-Based Education

▶   You have said that it is a paradox that our education system is not more people-oriented. What do you mean by that, and how can education be made more people-oriented?

"IT systems are created and used by people. For this reason we need courses in psychology to produce good IT managers and systems developers. In my view knowledge of how people assess things should, in particular, be included in the curriculum. Asking oneself questions like: When do I make mistakes? What does it take to create a system that is easy to use? There are human aspects to the vast majority of what we do. The only reason that we are very focused on mathematics and logic is that that is where we come from. We spring out of a mathematical mind-set."

▶   So that is our anchor?

"Yes, in a way mathematics is our anchor. The anchor is going to be the crux of your article, I see."

▶   Yes, that was where we started our conversation, so perhaps the anchor has become my anchor. Which of your studies are you yourself most satisfied with?

"My anchor studies have probably had the biggest impact. That a figure which pops up early in the proceedings – from goodness knows where – can have a disproportionately large influence on the entire process. The importance of using terms taken from psychology also in the world of IT. My contribution is not so very great, but I think it's this concept that has had the largest impact in practice. And taking a multi-disciplinary approach – it is by combining insights from several different fields that I have made my greatest contributions. That I find information in one place that can be used somewhere else. That's what I do mostly."

▶   So you are a bridge builder?

"Yes I am. Between different disciplines. My personal goal is to be so absorbed in what I'm doing that I am surprised every month when my payslip arrives. That's when I'm in the right place. And for me that's a large part of being happy."

We have come back to happiness again, and Professor Jørgensen talks about isolation and self-control. When we feel isolated, he explains, we are less inclined to exercise self-control. It is normal for people to cooperate, he goes on. It is when we *don't* do so that we end up on the news. The same goes for IT developments. They are normally successful. People are capable of producing the most amazing things – huge, complicated systems. And data security! If you think about how much is happening around us in the world there is surprisingly little misuse of data. The surprising thing is that misuse does not happen more often. It shows the degree of trust that is actually built into the systems with which we surround ourselves.

▶   Nevertheless – we could do better. What do you think is the way forward for IT projects? How will project management be improved, and what do you hope will happen in the time ahead?

"I hope that we will be able to assess vendors mostly on competence and less on price. It requires more of a customer, particularly in the role of purchaser. I also hope for greater psychological insight – into our rationality and our irrationality. That we become more rational with regard to our irrationality. Ha, that was good!"

▶   More rational for whose benefit?

"It's just a matter of efficiency. That we take the right decisions and make the projects as efficient as possible. That we are evidence-based in what we do. How many times during this conversation have you thought about the cake, by the way?"

▶   Not at all. That probably means I think this is absorbing. And you?

"I've looked at it a couple of times. So my training in self-control has been greater."

▶   But you may have got less out of the conversation. Our agreed round of criticism is about to start, I can tell… But before that – what is it that inspires you?

"The desire to understand what I have not previously understood. To read other people who have seen connections that I have not seen. I'm also inspired by the thought that, one fine day, I will carry out the perfect experiment. Both perfectly executed and with great impact on how people think. So far I'm taking tiny steps. My research is just taking tiny steps forward."

▶   Okay, it sounds as if you are ready for a round of criticism. I suggest that you start.

"First of all. I don't usually wrap criticism up with a positive preface. I don't think it's necessary, but here I feel the need to say that I very much like the fact that the interview has not all been one way. I liked that. Normally I don't start with that kind of positive comment."

▶   It's psychologically correct since… since it enables me to take the criticism. The brain is built such that we remember negative things much better than positive ones.

"Not my brain."

▶   In that case you're very lucky.

"I've always been like that. Objectively I may have done something really badly, and I am perfectly aware that I have missed the target entirely. But even so I remain totally convinced that I'm the world's best."

▶   So you are over-optimistic, and it works like a charm!

"I've discovered that not everyone is like that, and that I can't behave as if everyone else is like me. I have to take into consideration that some people may think I'm being brutal when the criticism starts. My criticism of you would most likely be that you don't have a lot of IT competence. You know that perfectly well yourself. So you have read up on it, and that's good. But I'm not 100 per cent confident that everything will come out right here. I'm prepared for you to make a lot of mistakes in what you write about the IT-side of things. Occasionally I felt that this could have gone better with a bit more structure about what we were supposed to get through. That we had thought through beforehand what the message should be here. That struck me. So we'll have to see what comes out of it. Proof of the pudding is what comes out of it. I am keenly anticipating how it turns out."

▶   Yes, you'll see. And the criticism hit the mark. I really feel it. Now it's your turn. And I'm now so cranky after what you said that I'll cut straight to the criticism… What surprises me is that you are not familiar with the Norwegian behavioural economics environment. That you don't know psychologist Per Espen Stoknes at the BI Norwegian Business School, economist Alexander Cappelen at the Norwegian School of Economics and Business Administration (NHH), or BI philosopher Øyvind Kvalnes. Together you would have made an amazing quartet! You could have achieved something big. It's a bit sad that you – an avowed bridge builder – are stuck in a computer science silo. You have to seek out people who are working on the same things as you, irrespective of whether you call it computer science, behavioural research or economics. The other thing is that it seems as though you don't completely believe in the psychology. Mathematics is so firmly established as an anchor within your field that most of the people you work with undoubtedly think you are a bit strange – an outsider who fiddles about with some psycho-babble stuff. What they don't see is that you represent the next step. They cling to their old anchor, while you are leaving it behind. It seems as though their anchor also affects you. You talk about research on happiness, which is also perfectly relevant in relation to the psychology of IT projects. At the same time you are constantly being pulled back to mathematics – as if it's that which remains the real field of study. In that way mathematics can act as a sea anchor for you.

"I don't entirely disagree with you. The emotional aspect is certainly much ignored. Anger, grief and similar emotions have not been fully included in the models."

▶    No, and it's a shame. For example, what do you think is the reason why the people in your experiments refuse to abandon the views you have planted in them, even after you've told them that they've been manipulated? Because they've become fond of their position. It means something to them, and having to give up the argument is akin to a bereavement – even though it has been planted! We get angry. And that's how conflicts start. And perhaps over-optimism as a kind of consolation?

"You may have a point there… We do have studies which show that as soon as we have selected something, that thing acquires an added value for us. It's interesting what you said about grief. It's not a word that's used in IT development and process improvement."

▶    Yes, it's used in business. Annicken Rød, an executive at the IT company Cisco, talks about the grief of transition. She sees it in connection with major changes in the workplace. There is a reason why seven out of ten mergers end in failure. There are so many sad people who have had their identities knocked out of kilter.

"You've given me something to think about there. I have taught on this subject, but I've not been very conscious about how I've phrased things. I use words like 'altered balance of power' and 'opposition to change', but have probably had a tendency to shy away from the emotional words. There's something in this. And now I have a suggestion. We divide the cake in two and call it even."

So speaks a true bridge builder.

# Through the Looking Glass into Digital Space

*An Interview with Paola Inverardi by Dana Mackenzie*

The question of how to make computer science more of a science was a recurring theme at the Challenges in Computing symposium. Martin Shepperd addressed that question in the context of machine learning and fault detection and Paola Inverardi talked about it in the context of software architecture. While Shepperd looked at reliability from the viewpoint of the software producer, Inverardi argued that the digital world is already moving to a new paradigm where the producer no longer plays the role of a guarantor of quality. In this world, a piece of code is simply an artefact that may work in the way it is intended or not. It is incumbent on the user to determine, by experimentation and observation, how the software behaves and whether (and how much) it can be trusted.

In the future, Inverardi said, 'Everyone in the world will be a user and everybody will be a producer'. New software artefacts will not be pieces of code, but combinations of services that are available in different places on the Internet or in the cloud. To make this possible, we will need integration mechanisms that allow different services to communicate with each other without special effort on the part of the user.

At the same time, Inverardi said, a new software development process will have to be invented that begins with identifying goals and then cycles through multiple iterations of a loop. The first step in the loop is to explore the digital space for resources that (perhaps) do what the user wants. The next step is to integrate them with other services, using the integration mechanism. The final step is to validate whether the new integrated software works. If so, then the developer can finish. If not, the developer goes back to the exploration or integration step, looking for new resources or trying to figure out how to better integrate the service.

According to Inverardi, the new process will require a radical change in perspective in the software engineering community. Software engineers will have to move from what she calls the creationist point of view, in which the producer or creator supplies all the information on the behaviour of its system, to an experimental view, where 'the more we observe, the more we know'. Instead of *verifying* that systems perform as expected, the software engineer will *validate* that they perform in the way they are required to. As the modelling of uncertainty becomes an explicit part of the process, software engineering will become more of a science.

As an example of the exploration step, Inverardi described a tool called StrawBerry, which she and her colleagues developed, that constructs an automaton that models the 'behaviour protocol' of a Web service based on data type analysis and testing. The automaton is then checked in a variety of ways to make sure it conforms to the behaviour of the real Web service.

Inverardi concluded her talk with two metaphors from children's literature. So far, she says, software engineers have been like Belle, trying to learn to trust the Beast of software. Instead, she exhorted the audience, 'Let's start playing Alice', and start exploring the wonderland of digital space.

The following interview was conducted on January 20, 2012.

A. Bruaset, A. Tveito (Eds.), *Conversations about Challenges in Computing*,
DOI 10.1007/978-3-319-00209-5_10, © Springer International Publishing Switzerland 2013

▶    How did you get into computer science?

I moved to Pisa to study computer science in 1975, when I was not quite 18. Pisa was the first university in Italy that had a *laurea* [bachelor's] degree in computer science in the faculty of science. I had no idea what computer science was and actually very few people did at that time. In Italy it was named *scienza della informazione*, the science of information. Many people in the general public thought that it was like journalism, because that was what they interpreted information to mean.

I had a set of very good professors, who were very young and who are now very famous. The person I worked with for my thesis for the *laurea* degree was Ugo Montanari and I had other professors, such as Giorgio Levi and Carlo Montangero. Most of these people came from Politecnico di Milano, so they were engineers in terms of their basic education. There were also some people from physics, but not many mathematicians. In the first two years the program was like a normal science degree, with a lot of mathematics and physics, along with just one introduction to programming course and in the second year something like algorithms. It was only in third and fourth years that we had more computer science classes such as programming languages, compilers, and theory.

My first experience in programming was with punch cards. Really! But you know what? I think it was very good to start this way, because we really understood what was going on from the very beginning, from digital circuits to a programming language. It was very valuable.

After I graduated I started working with Olivetti, back in 1981. This was also the year in which the first big national project in computer science started. It put together businesses such as Olivetti and almost all of the research groups in Italy that were devoted to computer science. The project had a number of different subprojects. The project I worked on, called CNET, was about building the whole chain from virtual machines to programming language to development environments in order to support distributed programming. At that time there was a strong interest in distributed programming and in distributed networks and machines, even though there was no Internet at all.

This was the Ada period. Ada was chosen as our programming language. Since the people in Pisa worked on formal semantics, we were following this path of building tools from formal semantic descriptions.

▶    What does formal mean in this context?

At that time the main focus was to talk about semantics as opposed to just syntax. While compiler theory was well developed, so that from the syntactic point of view there was nothing to discover any more, there was an emerging need to understand languages from the semantic point of view. One of the driving forces at the time was trying to standardize compilers to enhance the portability of programs. The big problem was that you could have programs written in the same language, but since they were compiled with different compilers on different machines, there was no semantic consistency between them. They could actually not be portable from one machine to another.

So there was a push in Europe to characterize programming languages from a semantic point of view. One of the strong contenders was the denotational approach of Dana Scott, who was at Oxford at that time. Then there was another thread of work starting at that time, built around the algebraic characterization of programming languages and data type specification. So we started talking about the specification of programs. There was this famous group, called the ADJ group [at IBM], who was promoting this view of using algebraic specifications in terms of signatures and axioms to characterize specifications of programs.

▶    How long were you at Olivetti?

For three years. After Olivetti I moved to a research institute called IEI-CNR, Istituto di Elaborazione dell'Informazione of the National Research Council, and I started working more on programs and tools. I went to the first conferences in software engineering. At the end of the 1980s I met Alexander Wolf of AT&T, who is now at Imperial College London. We started working together on software architecture, a new topic at that time. I had this formal background and he was a little more practically oriented. It was the beginning of my activity in the software engineering field, but he was already well connected with the software engineering community.

> ▶   What is the difference between software architecture and software engineering?

Software engineering is a general area that emerged at the end of the 1960s, with the awareness that the complexity of software was advancing, so a more engineered approach to the production of software was needed – as opposed to the more naïve way that software was developed at that time – to keep under control the final quality and the final properties of a software product.

Software architecture emerged from the recognition that when you build complex software systems, people were implicitly following some kind of reference way of putting subsystems together. A common example is the client–server model. This is a typical architecture. This had been to some extent implicit. At the beginning of the 1990s Alex Wolf and Dewayne Perry at AT&T and David Garland and Mary Shaw from Carnegie Mellon started talking about this characteristic as relevant and important to identify in an explicit way. This also supported the birth of component-based programming and systems.

> ▶   What does that mean?

Again, if you follow this path that software engineering tries to standardize tools and methodologies in the process of software production, then ideally you would like a standardized way to put things together with standardized, off-the-shelf components. Saying it's off the shelf means that the components come with their behaviour and qualities certified. I might add that this is still a vision that didn't really reach fruition.

> ▶   Is there any difference between that and what you were talking about at Oslo?

This is the historical trend. The next step after components was services.

A component is a piece of code that you can buy (ideally). You buy a component as a piece of code and put it together with the other pieces of code and then run the resulting system. But there are small problems to face, typically how to guarantee that the assembly of a set of individually correct components (usually a black box) does not produce misbehaviours.

The vision of services is not to put together the code but to reuse the service, reuse the functionality. If I need to produce something complex, I can ask for a service in a certain space.

Clearly this vision would not exist without the Internet, a digital world that has become more and more connected over time. The whole principle is that you can acquire a service that runs on another machine and gives you what you want. Of course, Web services are part of this vision.

For example, if you want to know the timetable of your train, you access one of the possible Web services that provide you with this answer. When you access this, you are not running the code; you are just accessing the service. The Internet makes this possible.

So now you can think of plugging together services, not code. This leads to a service-oriented engineering vision that is very popular in Europe, because the European Community has funded a lot of projects along this line. So now we talk about how to support choreographies, putting together services towards achieving your goal.

> ▶   How does this lead into uncertainty, the subject of your talk?

The point is that if you agree with this vision, now more and more we are surrounded by available services. These resources are no longer under our control. If I need a particular service, I can just go on the Internet and ask and I will probably get tens of possible services to use. What do I really know about the behaviours of these services? It will only be what I can experience about them.

The software world is becoming so complex, because it is full of elements that I can potentially use to do what I want and that I can observe. But my observation is theoretically limited. I will never be able to exactly know what this element will actually do when I run it. I will only have a certain confidence about it.

> ▶    Is that the idea behind the experimental worldview you outlined in your talk?

What I'm saying is that the way we will produce software in the future – in fact, we already do but not to the extent I envision – is more and more in this direction. This means that, as software engineers, we should provide tools and methodologies and processes that support this point of view. Supporting this view means that we need to abandon the world of creationism. That was exactly the world in which I was educated. In that worldview, software can fail because we have not programmed it well or because we have not verified it or because we have not declared explicitly what it does. This is the view of God, who is, in this case, the owner of the software. Once you do not own the software any more, this view doesn't help. I can only trust what I can observe. I might observe some crisp logical definition if it exists. But if it doesn't, I have to use other tools. This is what I meant by an experimental worldview.

I work in a faculty of science, so I have always been considered the black sheep. My colleagues say that computer science is not a science but a tool. However, the complexity of digital space is becoming so huge that it would compare with the complexity of the real world. When I think of my nephews, what is the distinction that they really make between the two? What do you think they will think is real? This is the change that amazes me most in my career. When I started computer science, it was clear what was real. It was perfectly clear what a computer was supposed to do. But the digital world, the potentiality, the virtualization of all our senses has created a paired world. Can we really distinguish between the two?

> ▶    How will it be different for your nephews and my nephews to live in this world? What possibilities will they have?

What I imagine is that my nephews will use these added resources to enhance their behaviour. This is something that I often think about: how little we understand how digital capabilities have changed the way kids learn. One easy example is memory. Probably you have been trained to memorize things and I have, for sure. Memory was our way of storing knowledge. My nephews are not trained to do that and they don't need to. For them, Google is a sort of add-on, something that enhances their behaviour and expands their memory in many possible ways. Does this mean that they are losing something? I don't know, because it's like saying that the fact that we use glasses has made our eyes lazy. It may be true, but would you go around without glasses? It's very difficult to say, but I think it's a very deep change.

Now my nephews know that this new kind of memory cannot be trusted. They know they have to search and not trust the first answer. You and I didn't know that. We were trained to remember and maybe we would have at home a big encyclopaedia, which was an oracle. They know that there is no oracle. The trustworthiness of the answers comes with observation. There are so many answers that they have to be selective and to decide what is right.

> ▶    It sounds a lot like learning to cope with uncertainty in the software services, as you were talking about earlier.

Yes. With this new process, this radically new way of thinking, we need to provide tools to observe and a different way of testing and validating what we have. For example, if I know that I cannot expect or achieve the level of confidence I would like to have, then I can work at the level of integration of the systems in order to mitigate the risks.

There are many techniques that have been used to cope with faults or uncertainty in specific fields where the creationist philosophy never could work 100 percent. These techniques have to be made available to the wider domain of software production.

►     **Can you give any examples of these types of techniques?**

The people who work in the area of fault tolerance have developed many techniques. They are all variations of redundancy, whether in the code or the component or the machines. Then there are techniques we can import from economics in terms of risk analysis. For big organizations, a software project can be a huge investment. It's important to make the right decisions at the right points. Economic theories are well suited for that domain. For example, in the early 1980s, Barry Boehm started the so-called value-based software engineering approach that is based on software economics. In recent years the approach has evolved by including also elements from cognitive science, finance, management science, behavioural sciences, and decision sciences.[1] There has been a lot of work on different approaches to testing, observation, and validation, depending on the kind of software and depending on what you can observe or cannot observe.

To give you an example, if you can only observe the output behaviour of a piece of software, you can only test and see how the software reacts to your stimuli. If the software inside, in order to provide you with that output, has to interact with other components, this is impossible to observe from your perspective. You would need to have access to a low-level machine perspective from which you can inspect the output doors of the component. In the real world, if you interact with me and you can only observe the interaction I have with you, then there is no way you can tell if I am actually getting my answers from somebody else. But if you can also hear noises in the room where you are with me, you can get a clue that something is happening. So in the digital world, if I can spy on traffic at the network level, I can see if there is traffic while I am questioning you and waiting for your answer. If there is traffic at your doors, your ports, I see that you are interacting with someone else. This means that I have a different level of observation. If I have that information, I can form a better idea of your behaviour and I might be better able to control or prevent misbehaviours.

Of course, if I can also observe your program's inner structure, your code, I can do even more. In the literature, there has been a lot written about observing what a code or components do, but there is no overall approach that actually takes as the driving force the level of uncertainty in the observations. This is what I was suggesting, to some extent mimicking what our nephews will do when they observe the output of blogs. Of course they have a goal, but they adjust, they integrate, and so on.

►     **Has there been any other technology that changed so much our sense of what was real?**

The only parallel that I can think of is books, the culture of written texts. It is through written texts that the generations before us have left their knowledge. In that way they increased the global knowledge, which is part of our real world now.

►     **Can you talk about some of the projects you have worked on and whether those have influenced your view of software architecture and the digital space we're moving towards?**

Yes. I will talk about two recent projects. For the last four or five years I've been involved in training exercises for the European Commission. From time to time they group together

---

1    See *Value-Based Software Engineering*, edited by Stefan Biffl, Aybuke Arum, Barry Boehm, Hakan Erdogmus, and Paul Grunbacher, Springer-Verlag, 2005.

experts in a certain area, put us in a room, and ask us to think about what the future will be in the next five to 10 years. This is a very difficult exercise in our domain, because five to 10 years is a huge amount of time. Nevertheless, it has given me a little bit of training in this direction.

The service vision and the vision of digital space come from the original vision of an Internet of Services, which has subsequently evolved to the Internet of Things. That includes things other than services. This vision of a digital space comes from the European vision. My vision is more at the software level, the level of software production, and it's basically this vision of the Internet of Services. The idea of capturing the uncertainty in the observations is my own vision. I cannot refer to a specific project, but there is a common understanding now that uncertainty is a key to this Internet of Services.

In order to be a little bit more concrete, we are working on a project that will end this year, in 2012, which is called Connect, a specific program of the European Commission that funds visionary projects in emerging technologies. The project starts from the assumption that it will be impossible in the future to provide standard solutions to all the possible interoperability problems that might arise in trying to combine existing systems. Until now, the answer to interoperability problems has been to set up a standard. But the more you see people developing software and introducing new protocols, the more you understand that the standard-making process cannot keep up. This is the starting point.

The project tries to address this problem at the middleware level and at the application level. This means that if we are given two (or more) systems that cannot interoperate because there are mismatches in their communication protocols, then the kind of solutions we are proposing will build a suitable connector on the fly if it is possible – an *ad hoc* translator.

If we have two systems that cannot communicate, then we assume that we can observe what they do. Then we use a machine learning approach enhanced with some kind of analysis tool. Based on these discovered protocols, we have a technique that works at the middleware level and up to the application level that tries to build a mediator, a piece of code that will be put in the middle to make the two systems communicate.

This is a very difficult problem, of course. We are able to provide solutions for a subset of these problems – not for all of them. You may see what we do in this project as a specific instance of what I said in Oslo, that we have to provide methods, tools, and support.

> ► It seems that your talk connects to several of the others in Oslo: Keith Marzullo's, Heinrich Stüttgen's, and maybe also Bashar Nuseibeh's. In Nuseibeh's talk, you have to discover not only how the programs work, but also how the people work with them.

I think they address similar problems at different levels. US Ignite is at the bottom level. We need something like US Ignite to connect the world. This is really the skeleton, the basic infrastructure we need. Cloud computing, the Internet of things, is at an upper level because it talks about coordinating available resources in such a way that they can provide their services to users. Bashar is talking a level above me, about how to formulate the goal so that I will express my wishes as a potential user in this immense digital space. He is trying to understand and specify what our nephews are looking for. Why are they searching? What do they want to achieve? I am in the middle, between Bashar and the real physical stuff on the ground level.

> ► What were your thoughts on the conference as a whole?

First of all, of course I knew Bashar and I knew about the Internet of Things. However, I was not really aware of how far the scientific computing and simulation community had gone. Natalya's talk was fascinating because of the immediate interplay between the digital world and the real world. It's not virtual duality; it's really merging the two worlds together. It's not simulation only. Simulation, in the end, is an engineering activity – what Natalya was talking about, with the use of simulation more or less coupled with the surgery. This real-time aspect is what makes it different.

▶   Does this give you any new ideas about the digital space that you were talking about?

Not really, because my domain is so different. I'm really interested in software in itself. To me, Natalya's talk provides more evidence of the fact that computer science is becoming an experimental science. It's getting difficult to distinguish between the two and it will become even more so in the future.

# Harmonizing the Babel of Voices

*An Interview with Martin Shepperd by Dana Mackenzie*

Modern-day computers and processors are too complicated for a human to understand. Software engineers frequently reuse and patch existing code rather than starting from scratch, but this eventually leads to the accretion of multi-generational code of great opacity whose behaviour no one can fully predict. One troubleshooting technique that software engineers often apply is machine learning: in essence, using machines to diagnose the faults in other machines. A machine learning program will test the response of a new piece of software to many different combinations of inputs and look for patterns in the results that suggest how the software may be flawed. It is really doing the same thing a human would do – learning from the past to predict the future – but a computer can try many more combinations and can detect more deeply hidden patterns than a human can. Fault diagnosis is not, of course, the only application of machine learning. There are many others, such as classification, prediction, and image and text processing.

In the past few decades, many different kinds of machine learning algorithms have been proposed: perceptrons, neural networks, decision trees, random forests, support vector machines, and Bayesian inference, to name a few. 'In the field of software defect prediction, there is no dominant method', said Martin Shepperd in his lecture at the Challenges in Computing conference. Each research group seems to have its own favourites. Although many studies have compared machine learning algorithms, they test them on different data sets and use different metrics to evaluate them.

Recently, Shepperd performed a meta-analysis to bring some clarity to this Babel of confusing and contradictory results. In the process, he made an astounding and in some ways very discouraging discovery. On a simple test of two-way classification, where the goal of a machine learning algorithm is to determine whether a particular piece of code is fault prone or not, 61 percent of the variance in the published results was attributable to the *research group that performed the test*. The actual choice of machine learning algorithm was nearly irrelevant! In other words, researcher bias overwhelmed any actual difference between the methods.

'Clearly we have a serious problem of bias in the field of machine learning', Shepperd concluded. And the problem may go farther than that. He quoted Stanford statistician David Donoho: 'Computer science cannot be elevated to a branch of the scientific method until it generates routinely verifiable knowledge'.

One solution to the problem of bias would be to practice simple precautions such as blinding (in particular, blind analysis), which are *de rigueur* in other fields of science but often not applied in studies of machine learning. Shepperd also suggested more cross-training between research groups so that they can develop expertise in different techniques in order to compare them more fairly. 'It does you no good to compare a good case-based reasoning algorithm to a bad Bayesian network', Shepperd said.

The following interview was conducted on January 30, 2012.

A. Bruaset, A. Tveito (Eds.), *Conversations about Challenges in Computing*,
DOI 10.1007/978-3-319-00209-5_9, © Springer International Publishing Switzerland 2013

▶     What got you interested in computing?

I did my bachelor's degree at the University of Exeter in the southwestern part of the UK, where I read economics. I quite enjoyed it and I nearly got a job as a research assistant for a group that did modelling of transportation infrastructure. But just before I started there was a change of government, back in 1979, and Margaret Thatcher became prime minister. She was not too keen on anything government funded, so I was made redundant before I even started!

At the time I had a bit of a motorcycle habit and I still do. Several of my motorbikes all went wrong at the same time, so I thought that I'd better get a job. I saw an advert for a computer programmer and I thought I'd do it a few months and then get a proper job. It turned out that I was just in the right place at the right time, so I got to do some challenging things for someone quite new to the field. This included the opportunity to work on novel multiprocessor architectures from microcode on up. It involved coding in hexadecimal up to high-level languages such as Algol 68. The project also incorporated some document processing coding and work on high-speed (well, high speed for those days) data protocols that had to be extremely reliable.

It was an exciting time because relatively few people had formal training in computer science. We just discovered a lot of the principles of computer science by trial and error. If you had the curiosity, you could do some pretty good stuff.

After working there for four years instead of the four months I'd originally been thinking, I started thinking that if I could get off this treadmill, I could clean up this library and I could figure out how to do that testing properly. I needed a chance to reflect, so I decided to go back to university. I earned my master's at Aston University in Birmingham and my doctorate at the Open University, also in the UK. After that I was a lecturer at Wolverhampton Polytechnic, which is now Wolverhampton University. Then I moved to Bournemouth University and finally moved six years ago to Brunel.

▶     So after getting back into academia, you never left!

I sort of drifted into an academic career, mostly because it gave me the luxury of time to reflect, which I hugely value. That is what led me to look at issues having to do with prediction.

The major approach to build what are called prediction systems is to try to learn from the past. It's an algorithmic way of reflecting, if you like, a semi-automated way. A lot of the techniques I specialize in, such as case-based reasoning, are based on the premise that history repeats itself, but with a lovely little rider that it never repeats exactly. Therein, I think, lay much of the challenge and much of the interest. If history repeated itself exactly, prediction would be trivial.

▶     How do prediction systems differ from machine learning and from artificial intelligence?

Machine learning is the technique of automated inductive reasoning. It's one of the larger topics within computer science. I would say that artificial intelligence has a tendency to be more philosophical and less applied than machine learning. Some of the artificial intelligence techniques are more deductive. You might say here are the rules and principles and ask what new things can we deduce from them. In some sense it's like algebras or axiomatic systems. You start with something given and try to deduce new things. Machine learning works in the exact opposite direction, from the particular to the general instead of from the general to the particular.

To ask which one is better would be a foolish question. In many ways they are complementary. I don't think anyone in machine learning wants to avoid deductive reasoning. But there are classes of problems where we don't have the kind of fundamental tenets that would be starting points for a deductive program.

Software engineering is a good case in point. Software systems these days are mind bogglingly complex. They are far beyond the intellectual span of an individual. That was beginning to happen even in my day. We had systems in three or four languages and probably 1 million or 2 million lines of code hosted on a dozen different kinds of computers. That would have been one of the more complex systems in existence. Nowadays, it would hardly provoke comment. Plus, you have an amalgam of ideas generated by people with different backgrounds and different thoughts on how you should solve problems, all melded together over many years. The memory of what they did is lost because people move on.

These systems are some of the most complex artefacts that civilization has yet developed. That fact might not be obvious to everyone because they are intangible and, in addition, when they are working, you don't notice them. A modern car may have upwards of 50 different processors onboard, developed by different people with different architectures and different processes, and they need to work to solve quite complex tasks in a coherent manner. We don't have three or four simple rules of software engineering that apply to such systems. There are no laws. Informally, people talk about laws when they think they've observed a principle that seems to be widely applicable, but it's not a law in the sense of natural sciences.

> ▷  If a computer simply follows instructions in a deterministic way, why can't
>      you predict its behaviour?

If you have a small task, you can probably produce a program that satisfies your requirements with a reasonably high probability of success. But these things are hard to demonstrate. But for reasonably large software, exhaustive testing is impossible. It's a combinatorial problem. You may have 10 inputs that are integers that are bounded between the smallest and largest that the machine can represent and you add to these a few Boolean variables and input strings. When you put all of these together, the space of possible inputs is vast. Exhaustive testing isn't possible, so usually people divide it into classes. You might believe all negative integers are the same as far as your program is concerned, so you just pick one number. That's a fair heuristic, but it doesn't guarantee the code is correct.

Back in the 1980s and 1990s, people tried to create formal mathematical models of software behaviour. The hope was that we could prove some properties we want to guarantee. For instance, if the software is something to do with networks, you might want to show that the system won't deadlock. Using a certain refinement process, you'd take your mathematical model, defined in terms of sets or mappings of functions, and you would try to refine that into your code. At each step you'd prove that the refinement satisfies its specification. In that way you would prove that your code is correct.

First problem: Practically, that is not feasible. Second, things that are easy to specify in a formal sense are often the least interesting properties. Then even if you've proved that your code is correct, you still have to prove your compiler is correct. Even after you've done your refinements into source code, what kind of object code does your compiler generate? C is quite an unsafe language that can do lots of weird and wonderful things you didn't expect. And a lot of the temporal properties are difficult to specify. Finally, on top of all this, there are issues of usability. If your code is for some kind of Web-based system or e-commerce system and it takes too long to execute, then it's utterly unusable. You'd still have to reject that, even if in some narrow sense it's correct.

For all these reasons, I think people backed off quite a lot from that kind of approach. You'll still see it as an adjunct to model-driven approaches where your understanding of the domain is encapsulated in a set of relatively simple models, but proving the program correct for anything that's not trivial is not going to happen. It would be most valuable to have that proof of correctness for the really important, complex software systems, but those are precisely the ones that are the least tractable.

Another difficulty is this weird idea of software that ages. You might wonder, how can code age? It makes no sense, because code is what it is. But imagine the accumulation of 1000 different patches, 999 of those made by people who did not originally develop the software, altering the original code as little as possible. They put endless loops and branches around things, rather than doing any rewriting. So any structure there was in the original code is lost. The documentation becomes very unreliable, so all you can trust is the code itself. What's the point in spending days looking at a document that probably isn't right?

▶    **Can you give any examples of software that didn't behave as it was expected to?**

A famous example is the failure of the Ariane 5 rocket, because it's extremely well documented. [Note: This same example is discussed at length in the interview with Bashar Nuseibeh. For this reason, the details are omitted here.] This was an example where a number of seemingly small, perfectly reasonable decisions were made to reuse code from the previous spacecraft, Ariane 4. The engineers thought what's the point of testing something when you've already used it successfully? The whole chain of decisions came together to produce a software system that failed.

▶    **Do similar things happen in less critical programs all the time?**

That's the most spectacular one, but software failure is a relatively common occurrence. Take telephony, for example. You try to place a call but it doesn't connect. You think, 'I must have misdialed'. Generally you don't realize that it's a software failure. You just dial again.

Instead of trying to drive all the faults out, one of the tactics people use is to make the systems resilient. You admit that faults are inevitable and try to accommodate them the best you can and try to degrade the system gracefully in case of failures. That's why major outages for telephone networks are rare but they do happen and the telcos invest a huge amount of money and effort into trying to produce as reliable software as possible.

▶    **In your lecture, you talked about the lack of replicability of results in machine learning. You performed a meta-analysis of 200 papers. Could you take me through how you found those papers and what was the objective of the study?**

Let me roll back fractionally. Given the difficulties of formally showing the correctness of software systems, people turned to machine learning techniques and asked whether we could use them to predict where the problems are in an inductive way. If you could do that successfully, it would have huge economic importance. But machine learning is a pretty broad church, so there are many different kinds of algorithms. Because the approach is inductive and there's a lack of deep theory, it's a very experimental discipline. For example, we don't have much theory on the relationship between the structure of our data and how we would expect our learner to perform. Therefore researchers conduct lots of experiments.

The starting point for our study is that no one approach currently dominates the field. To be honest, I'd say it's worse than that. The field is kind of chaotic. There is a huge amount of activity and a lot of noise. So the natural next step is to say we don't need any more primary studies, because we already have plenty of them, but we need some more sense making.

Some colleagues of mine conducted a systematic literature review. It's a similar idea to the practice in medicine and clinical science. The idea was to do a very rigorous objective review where the idea is to be inclusive. Having determined your research question and inclusion criteria, the goal is to include every single study instead of cherry picking. That is a lot easier these days with ScienceDirect, Scopus, and Google Scholar, but it's still a lot of work. My colleagues conducted this review and they located 208 studies. These were not the only 208 that had been done, but they have to pass certain inclusion criteria. For example, they have to report sufficient detail and be tested on real-world data. The inclusion criteria have to be set at a certain level so that you're not just looking at early drafts and opinion pieces. They have to be proper, refereed studies.

Even at this stage, my colleagues were struggling. It started to become very clear from their analysis that an informal approach dominates in this field. There are so many design decisions in the way that people conducted their primary studies and so many customized algorithms. I suggested doing a kind of meta-analysis. Instead of asking which technique is better, let's ask what factors affect the variability in their results. Assuming we can in some way assess predictive performance, we can do a rather complicated analysis of variance to see what factors explain the variability in results from study to study. If people are doing good science, we would expect the choice of machine learning algorithms to be the most important factor.

From some simulation work I've done a few years ago, I expected that there might also be some interaction between the structure of the data and the performance of the particular machine learner. It might be that some algorithms might not deal as well with missing items or we might have problems with correlations between explanatory variables or some techniques may deal better with outliers than others. So the second factor was to look at the type of data they use. Third, there's some discussion on what types of metric or measure you might use. Should you try to look at the process or look at the amount of time or should you focus on how the software has evolved? When something changes a lot from one release to the next, that clearly has some significance, so you might want the machine learner to focus on that kind of information to form a basis for some prediction system.

Finally, because I had been reading some articles from the social sciences about researcher bias, I thought we should put in another explanatory variable, which is the research group that does the work. To operationalize that concept, we ran a clustering algorithm based on the idea of authorship. Every time two researchers co-author a paper, they get clustered together. We found clusters ranging in size from one singleton, who only publishes by himself, up to groups of eight to 10 people.

When I did this, I was not expecting the researcher bias to be that big. For sure I thought the choice of algorithm would be significant and it would have interesting interactions with the data set and possibly even the research group. Perhaps there will be some expertise issues: One group might be really good when they use a particular algorithm, another less so. But in fact, the research group turned out to be by far and away the single most important thing that you need to know.

Essentially we examined four factors to try to explain the variability in prediction quality results from each study. These four factors were choice of algorithm, dataset, type of input metric, and research group. We also had to standardize prediction quality as a single, comparable response variable, that is, the thing we wanted our model to explain. In practice, researchers use many different measures. In the end, I used a Matthews correlation coefficient, as it uses all the confusion matrix[1] and is easy to interpret. (It ranges from -1 to +1, with zero indicating no relationship, in other words, the prediction system is just guessing.)

I was also surprised that the analysis didn't have to go very deep. We just looked at linear random effects models, nothing exotic. In part I took this approach because I believe there is sometimes a danger of rushing into overly complex models which may not be warranted by the quality of the data or depth of understanding. We allowed for interactions between factors. We didn't demand that the factors be orthogonal because we didn't think they would be.

I was quite surprised at the goodness of fit of the entire model. The way we coded for the different factors was very crude and yet we got an extremely good fit. In general, if you get a fit quite a lot better than 80 percent, you think you're onto something. Fit was assessed in the usual way, using the coefficient of determination (taken from the sum of squares of the residuals).

Another quite important decision was that we used a random effects model rather than a fixed-level model. What that means is that I don't care whether one machine learning algorithm does better than another one; I just want to know what things cause the predictive performance to vary. I don't think we're doing good enough science at the moment to reason out what are the best ways of machine learning. It's all to do with the replicability of our work. If we can't replicate, I don't think we're doing science.

---

1   The confusion matrix is the $2 \times 2$ matrix of true positives, true negatives, false positives, and false negatives. As Shepperd pointed out, focusing on only part of this matrix can lead to very misleading results. If 95 percent of the software being tested works correctly, then a fault detection algorithm that certifies everything as correct will give true results 95 percent of the time. Nevertheless, it will be absolutely useless!

Incidentally, we've just upped the number of studies to 600 by relaxing our inclusion criteria a little bit, because some of my colleagues commented on this. We find it's a very stable relationship and so we feel even more confident in our results.

► **What do you see as the way ahead? Do you have any thoughts about how studies can be made more replicable?**

It's very easy to see this as a negative result. But if we interpret it right, it tells us how we should be doing our research. Otherwise we're taking government grants and pushing out all of these primary studies that, to be brutal, aren't helpful.

There are a number of things we need to look at. One is reporting protocols. These should include all of the relevant information, the subtleties to algorithms, the way that data are pre-processed. This information is not easily accessible now, sometimes perhaps because of page limits. The devil really is in the details.

I think the other thing I'd look at, apart from reporting protocols, is how we deal with research expertise. There are parallels in the medical world, with complex surgical procedures. Other research teams have sometimes been unable to replicate results until they sent people to the first research centre because there's an expertise education issue. There are two ways you might handle that. You need more inter-group collaboration, so that people who are good at one technique work with people who are good at something else. We also need to look for ways of raising everybody's expertise.

For example, let's suppose it's a study done by my group. We have done a lot of work with case-based reasoners, so we're not bad at that. We might want to compare them with a different algorithm we have modest knowledge of, such as support vector machines. For that algorithm, we might just take the bog-standard settings because we don't know any better. We're in danger of using one technique in a very sophisticated way, while we're using another technique in a very naïve and dumb way.

Although we're looking at this particular branch of machine learning, everything I'm saying is more widely applicable. We need to share our data. A lot of these algorithms are very complex. It's not just a case where you can turn a handle and get a number out. There may be a few rules of thumb and if you're not an expert in that area, you might not know them. The way some of this research is done, to be honest, is to put a graduate student in the basement for two years and tell them to optimize this algorithm somehow. After two years you let them out and they get a result a little better than someone else's and you publish it! The process is invisible to anyone else and also it's very costly if there's a lot of trial and error.

► **Has your meta-analysis been published?**

It's under review but, unfortunately, it's not published yet. We've had some dialogue with referees for one journal, where they are concerned about the wrong things. It's dragging the process out a bit.

► **Is this paper going to make you unpopular with other computer scientists? I wonder if the referees are using these technical issues to hold up the paper?**

I suppose it's possible, but I hope not. If the work is scholarly, it will be published; if not, it should be improved. This article isn't finger-pointing. My colleagues and I have done primary studies in this area too and we're as implicated as anybody.

► **Since you brought up the medical parallel, have some fields successfully coped with these issues and, if so, can you import some methods from them?**

Exactly, I think that is one of the things to look at. Then there's another very interesting question. I'd like to look at a completely different problem domain, such as gene expression data, to see if these issues are more widespread. I rather suspect they will be. Then we might look more generally at the extent of researcher bias in computer science and beyond. Two years ago, I would not have predicted I would be thinking in such a fundamental way about how to do science.

►     Finally, what did you think of the other talks at Oslo?

I was very inspired by some of the large-scale computer science that was being done. For instance, there were two talks about the modelling of the Earth's mantle and plate tectonics. Also, the cardiac modelling was very impressive; so much computer science underpins it.

    Without being overconfident, I think we're beginning to see that computer science is coming of age. I know that feeling is somewhat at variance with my earlier comments about machine learning, but we are doing big science and truly interdisciplinary work. That's encouraging. If this event had been held even 10 years ago, I don't think we would have the same sense that we're doing big models and important stuff. I don't want to come across as being complacent, but I was quite heartened by the whole thing.

# Mediating between Man and Machine

*An Interview with Bashar Nuseibeh by Dana Mackenzie*

People are illogical. They say they want one thing when they really want another. Or they say they want one thing when they *need* another. And often, especially in the context of privacy, they can't actually verbalize what they want. They simply know it when they have it.

Computer programs, on the other hand, are inherently logical. They are (at least traditionally) based on concepts of Boolean logic. If condition A holds, the program takes action B. If not, then it takes action C. A program typically does not take into account variables that it is not told to take into account: what the weather is like today, what mood the user is in, where the user is physically located.

Mediating between these two worlds is Bashar Nuseibeh's specialty. As one of the early researchers in the field of requirements engineering, he has found ingenious ways to tease out the 'voice of the customer', something that is often hard to do and cannot always be done by simply asking questions. And then he helps software engineers deal with the users' messy, sometimes self-contradictory expectations. At times this forces him to sacrifice some traditional views of software development, such as the waterfall model, in which a requirements document is drawn up at the outset and then the software is designed and implemented. He argues that in the real world software design very often needs to be an iterative process where requirements are modified again and again.

At the Challenges in Computing conference, Nuseibeh described the challenges involved in understanding users' notions of privacy. First, he pointed out, the concept of privacy is very ill-defined. He cited several different definitions, from the 'right to be left alone' (Supreme Court Justice Louis Brandeis) to 'the right to control … information about them[selves] and decide when, how, and to what extent information is communicated to others' (Alan Westin). The latter definition in particular suggests the very nuanced nature of the 'privacy' concept. It can change depending on the time and context and, as Nuseibeh said in his lecture, 'It is not just about *hiding* information but also about *providing* it'.

The next difficulty with understanding users' privacy requirements is the difficulty of getting them to articulate them or even to remember accurately what they were doing when privacy issues arose. Nuseibeh cited a study his team published in 2009 of user behaviour with a mobile Facebook app. The goal of the study was to understand in a qualitative way how users decided when to update their status, write on someone's wall, upload a photo, and similar actions. Did they do it more often when they were alone or with other people? Did they make any attempt to hide their actions?

Perhaps the most remarkable thing about the study was not the results but the methodology. Participants in the study were asked to fill out a very short online multiple-choice questionnaire about their action and then to choose a trigger phrase to help them remember it later. Nuseibeh's team found that the participants were able to remember actions and their contexts in considerable detail in a follow-up interview, even weeks later, but

A. Bruaset, A. Tveito (Eds.), *Conversations about Challenges in Computing*,
DOI 10.1007/978-3-319-00209-5_12, © Springer International Publishing Switzerland 2013

only when they were reminded of the trigger phrase they had chosen. Without the trigger phrase, they couldn't remember where they had been when they uploaded a photo, for instance.

In another experiment involving privacy, Nuseibeh played two videos of a fictitious product, a pair of glasses for dieters that will monitor what food you are looking at and tell you how many calories it contains. One video showed the new product in a favourable light and the other one showed it negatively. 'The contrasting visions elicited a wider range of results', Nuseibeh said, including some the researchers never thought of. Some people said they were worried about what the glasses would tell them if they looked at another person!

Finally, the counterintuitive nature of user behaviour was illustrated by a third study, in which Nuseibeh's team followed the usage of a 'buddy tracking' application within families. 'The more controls we gave them, the less they used them', Nuseibeh said. Apparently the additional capabilities made the trackers – not the trackees! – feel more uncomfortable. 'This shows that privacy is very personal and very individual', he concluded. Software engineers who are not aware of this fact may find their products being used (or not) in unintended ways.

Nuseibeh concluded the talk with his list of challenges for the future, which includes questions such as these:

- How do we incorporate very qualitative models into engineering?
- How do we determine privacy requirements from qualitative data?
- How do we relate models of user behaviour to software architecture?

While our interview does not answer these questions, it does provide some insight into the mind of a scientist who has made a career of studying the quirkiest part of any computer system: the user. The interview was conducted on December 30, 2011.

▶    Your biography says that you came from Palestine. How did you end up in the UK?

I was born in June 1967, a time of conflict in the Middle East. I am originally from Jerusalem but I lived the first few years of my life in Jordan. My family then moved to Edinburgh, Scotland, for a couple years, where my father specialized in paediatrics. Then we returned to Jordan, where I completed my primary school years, and then left for Abu Dhabi for my secondary schooling. I started as an undergraduate in the UK in 1985. I've been in the UK ever since, except for the first few years after finishing my master's, in 1989.

▶    How did you get started in software engineering?

It started when I did my undergraduate degree on a topic I didn't really understand until after I had finished it. I was really interested in computers, everything from Casio-type one-line screen computers to Apple IIs. I knew I liked software, but I didn't know what you could do if you wanted to study it at university. I knew that in the Middle East people were quite keen for someone to be an engineer. So I found a course called computer science in the School of Engineering at Sussex University and figured that would give me the respectability of an engineering background as well as provide my computer science. But it turned out to be pretty much electronics and control engineering, designing circuit boards, and software played a very minor part. I got depressed because I wasn't doing much software.

Then I tried to do a masters degree that was more focused on software. At Imperial College London, they had a master's program called FAIT (Foundations of Advanced Information Technology). But it turned out that Imperial was focusing more on the word *foundations* than *information technology*. I spent a year studying logic and still didn't do any software.

But at that time I met somebody who probably has had the most influence on me professionally, Anthony Finkelstein, who taught the only course with *software* in the title. I enjoyed what he was teaching and decided to do my master's project with him. That was when I started understanding what software engineering might be about. He was the one who really had the perspective of the customer or user as the first class entity. That's where I got my inspiration and my guidance. I still didn't know what I wanted to do with my master's, which is why I went back to Abu Dhabi and worked for Tektronix, a company that made medical and electronic instruments. I soon discovered that working in industry wasn't what I wanted to do, so I contacted Finkelstein with the idea of doing a PhD.

What attracted me at the time was a project called SEED, Software Engineering and Engineering Design. The idea was to look at the relationship between software engineering and traditional [non-software] engineering design. It so happened that they were looking at electronic instrument design, which I'd had some experience of at Tektronix. At this point I started thinking about what the inputs to those two engineering disciplines were. The inputs were typically requirements, but it seemed that electronic engineers seemed to know what their requirements were much better than software engineers did. I wanted to understand why and how we could change it so that software engineers could have better input into their engineering and design process.

I ended up working on three projects during my PhD, which I finished around the end of 1994. It was a relatively quick doctorate, given that I was doing it part time, but I had a very supportive supervisor. My conclusion was that it was very hard to come up with single requirements for a lot of software systems. There are always many people involved, trying to tell you what they want. I came up with this notion of a 'viewpoints' framework that allows multiple partial descriptions of your requirement. Until then this had not been the accepted wisdom; you typically had one requirements document. I was advocating lots of partial requirements documents with links that described the semantics of the relationships between them. One of the interesting ideas was to tolerate inconsistency, to allow multiple views that were not necessarily consistent at the time they were expressed but were important to capture because that is what the customers were saying at the time. If you try to hide these views, they would forever disappear from the documents you were producing.

The difficulty was that as soon as you start moving to designs and algorithms, inconsistency is not a good thing to have in your documents. But requirements are necessarily inconsistent and partial. I spent some of my PhD and postdoc time playing around with logics and descriptions that tolerated inconsistency and allowed you to continue some reasoning and make some useful deductions even if you had an 'A and not-A' in the requirements description.

▶  **That's really going beyond traditional math and computer science. Was there any pre-existing work in that discipline that you could draw on?**

There's a long history of paraconsistent logics that tolerate inconsistency. A lot of people have thought about different ways in which you can represent inconsistency or uncertainty, such as probabilistic logics or Bayesian logics. In some ways this discipline was not so widely accepted and not so respectable. I came across a paper by a professor in my department, Dov Gabbay, with his PhD student, Anthony Hunter, called 'Making Inconsistency Respectable'. On the applied side there was a researcher named Robert Balzer in the US, building systems that allowed you to successfully compile programs even though they had errors in them. That paper was called 'Tolerating Inconsistency'.

So I discovered this body of very scattered papers on managing inconsistency in other ways than resolution. This was exactly what I was looking for, a way to express these multiple partial views so that you could ask stakeholders more questions. Inconsistency implied some sort of action; it was a useful driver for the development process. Without it we could not keep asking questions that allowed us to elicit or capture more information. Obviously at some point you needed to make choices about the inconsistent things that you wanted to build, but at least you kept a rationale of why you made those particular decisions. Those were my PhD years and soon afterwards. Even to this day, this is the work I'm probably most known for, multiple views and managing inconsistency.

This is very much a *post hoc* observation, but I guess my interest in conflict came from living in the Middle East. You had to live with opposing viewpoints and the idea was how to make progress in spite of them. I spoke with another logician here at Imperial, Bob

Kowalski, who is famous for developing an area called logic programming. As part of his own research, he had formalized the British Nationality Act. We had subsequent discussions about the Middle East conflicts and how you could formalize them. I don't think we made progress on either the formalization or the conflict, but these approaches were very much inspired by the real world, in which people were communicating their needs and requirements and describing their world.

There had always been this assumption that all the requirements are somehow stored in a single database and that you should strive to make everything in it consistent. That was the prevailing view, but not the view that Anthony Finkelstein and I thought should be the case.

▶  **When did you start doing consulting for companies outside academia?**

In many ways these things landed in my lap, because software engineering was an area where people felt there was expertise missing and they just didn't have the basic processes for understanding what systems to build. Again I think I have my supervisor to credit for that, because he got me involved in work he was doing with Philips. It still had the flavour of being input into my research. I almost never do consultancy purely to apply software engineering; it's always because there is an interesting research issue that I'm trying to investigate. With Philips, we were looking at an issue called requirements traceability. If requirements change or a product changes, how can you trace forwards and backwards between the requirements and the design?

▶  **What is your favourite or the most interesting consultancy that you ever did?**

I have to say NASA. It was part of my sabbatical. I was at this facility in West Virginia called the Independent Validation and Verification Facility, IV&V. I was surrounded by people whose sole purpose in life was to try to break software. They weren't involved in its development. They took software developed by one part of the organization, tried to check it against the requirements, and tried to break it.

Until that time I had thought my PhD was a little bit theoretical, in the sense that people weren't writing large requirements documents. But at NASA they were. At the same time, I realized that they were not *using* requirements documents, they were *abusing* them. The documents never reflected the real state of the software system that was being built. They were often fixed after the software was built or modified. It wasn't the traditional process, where you specify requirements and then build the system. That is called the waterfall model: first requirements, then design, then testing. What was happening at NASA was much more iterative and going backwards and forwards.

This was the first time I'd seen requirements in action on a very large scale. There were rooms full of requirements documents. They had documents for the International Space Station with multiple versions over many, many years. Understanding consistency suddenly took on a very different meaning. Consistency between what and what? What does it mean for one version to be consistent or to fix an inconsistency?

▶  **Do you feel your project changed the way things are done at NASA?**

Not at an organizational level. But remember, I was on sabbatical. My role was not to exercise any kind of change management. But it did lead to other things. My colleague, Steve Easterbrook, who was Anthony Finkelstein's first PhD student, was also working there and he was hitting his head against a brick wall, trying to change things.

One of the things Steve and I decided to do while I was there was to write a book, because we felt there was no good book that described what requirements engineering was. We spent a considerable amount of time coming up with a table of contents and a proposal. Unfortunately the book never got written. We had some chapters on the Web and it became a standard joke that it was still being written. One thing we did do was convert the table of contents into a paper, a requirements engineering roadmap, which, as it happens, is one of the most widely cited documents in the field. It's probably my most highly cited paper. It was published in 2000 at the International Conference on Software Engineering.

> Is that your favourite research paper or do you have others?

I don't think so, because it's just a review paper. I'm very proud of it, but the paper on multiple viewpoints, published in 1993, is the one I'm most proud of because 10 years later it received a most influential paper award. I've written one or two other very small papers that I really like that have had a little bit of influence.

For instance, I wrote a position paper years ago called 'Weaving Together Requirements and Architectures', which argued that requirements and software architectures were equal starting points in the software development process. The choice of requirements and architecture can happen simultaneously and you iteratively refine both as you explore problems and solutions.

In that paper I drew a sketch of two triangles, side by side, one labelled *requirements* and one labelled *architecture*, with a spiral line going from top to bottom. The triangle represented increasing knowledge of the requirements and architecture and the spiral was an indicator of going back and forth between the two triangles as you understand the influence of one on the other. I was probably not very sober at the time and so I called the diagram Twin Peaks. Now that is one of the things that I'm known for: the Twin Peaks sketch! There is even going to be the first international workshop on the Twin Peaks next year in September in Chicago!

There is little technical content in that diagram, but it must have struck a nerve because the requirements community and design community had been fighting for a long time over which was a more important element of the design process. My point was still that you have to keep dipping into the problem world in order to get technical solutions.

Security and privacy have really brought this out more than any other type of requirement that I've worked on. The problem world of malicious attackers and unexpected events is very characteristic of security problems. You can't specify everything in advance. The ability to specify requirements iteratively is very important, because if you take it to its natural conclusion, you might have systems that are constantly looking at the problem world and trying to adapt themselves in response to changes in their environment. We're moving from a traditional process of building a software requirements document, design, and solution to a world of more autonomous adaptive systems that are able to understand the world around them.

If you have a good understanding of the relationship between your models of the world and the models of your machine, then you can build it into the software. The machine can see if its model of requirements is up-to-date and, if not, it seeks to adapt and evolve that so that the system is reactive to new threats and new scenarios and events.

Although I'm interested in security and privacy, I'm more interested in requirements and the role that they play. I think they play a different role in adaptive systems and systems where security is important. They need to be a first-class citizen in the system itself and not just the development process.

> How did you get interested in these questions of privacy and security?

I changed jobs at the end of 1999 because I felt I was getting very comfortable at Imperial College. It was a very nice place, I had a secure job, and my career was going really well. I tend to get a little bit bored and it made me a little restless to feel that I could potentially spend the rest of my life there. I had tenure and everything was going really well, so I thought this was the best time to move. It's a bit strange, I know, but that is how I was thinking at the time.

At the time there was a potential opening at the Open University, a distance education university, which is very well respected and loved by the general public because of its social mission of trying to educate people who might not have had an education before or wanted to change careers or return from a career break. I wanted to explore how you could build a research programme at a university without undergraduate students on a campus. In 2000 the Open University offered me a full professorship, so I took the step and moved. I then spent the next 10 years developing the department and my own research agenda. At the time I felt I needed to pick something different to do.

While still working in requirements, I wrote that Twin Peaks paper and another paper on anti-requirements, the requirements of malicious users. I thought that anti-requirements were an interesting concept. Requirements are sometimes inaccessible not because the customer didn't tell you, but because you don't have access to the customer, which is the attacker. I wanted to see if you could potentially investigate security requirements before you build the system. Typically you build the system first and investigate if it's robust against certain attacks. I wanted to see if you could have a discourse about such attacks at the requirements stage. Can you have a requirements and an anti-requirements description together?

This is where the consistency issue came into play again. A malicious user might want to steal some money, break some secure mechanism, or access some assets. If the requirements say you must not be able to do this but at the same time you are able to identify some description that allows the anti-requirement to hold, that turns out to be a description of the vulnerability. This got me interested in security as an area of research because of this lack of access to stakeholders and because of this concept of negative requirements. Security requirements were requirements with lots of nots. Given that I had looked at a lot of A-and-not-A's in my inconsistency management career, I thought this was an interesting example of inconsistency management.

I got a little UK award at the time called the Leverhulme Prize, a nice little pot of money for people under 35 who've done something useful. That funded my initial foray into security.

▶    What sorts of problems in security have you worked on?

I spent a lot of time trying to understand trust. I couldn't quite get a definition I liked, although it's a concept that has been around in philosophy for a long time. People's trust behaviour is very different from what they tell you. They say they trust the water, but they only buy bottled drinks. I started wondering, how does this relate to notions of security?

I brainstormed on this for a while with a PhD student of mine and we came up with the notion that some of the most problematic requirements are assumptions. You assume that something will behave in certain ways and the system fails because people's behaviour is different from the assumption. Trust is very often an assumption about the system. You trust that no one will enter your office and you leave it unlocked. Or you trust someone will access a resource in a certain way through a computer system but you are mistaken. We decided that we would model trust assumptions as part of the requirements description. You surround the system with trust assumptions and you question them systematically. We constructed this entire framework that had trust assumptions at its heart and using argumentation as an analysis technique. It gave us more of an operational definition of trust, rather than something you could compute or philosophize about.

▶    It seems as if you're constantly trying to uncover or make explicit things
     that people have almost been deliberately not looking at. Are there any
     secrets as to how you do that?

If there are, they're secret to me, too!

I don't like thinking along traditional lines. I like to come up with something different from the usual arsenal. This has positive and negative points. Discussions with me are rarely focused because I always come up with things that are outside the agenda. One thing that characterizes my work is that it's often different from the prevailing wisdom or practice. I find that enjoyable.

I take the notion that publications ought to have something novel in them quite seriously, maybe sometimes to the detriment of depth. I'm lucky now that I have a large group, with people to help me work through the details that weren't worked through before.

Maybe the secret is to enjoy doing something different and to be willing to read outside your area. I go to conferences that are not traditional in my field. For example, I go to an annual workshop on security in human behaviour and to human–computer interaction conferences, as well as my bread and butter conferences on software engineering and requirements engineering. That's one reason why I travel so much.

▶ How difficult is it to persuade software developers that they need to take into account human behaviour and human–computer interaction? Is that a case that's hard to make?

It's getting easier but it's still hard. It's getting easier in the sense that people recognize that systems fail for social and economic and political reasons, not just for technical reasons. Most sensible companies and people will acknowledge that. The difficulty is that it's not obvious how to translate that knowledge into something practical. The other problem is that the software engineering community is very conservative. I find it almost impossible to publish my work in a traditional software engineering conference. I'm probably one of the most well-connected people in the field, yet I have difficulty persuading them that my research fits into their conference.

Fortunately, there is a lot of funding for multidisciplinary research if you can articulate it properly. But it's much harder to publish. If you're a young faculty member or researcher, it's a little risky to say, 'I want to work in this bridging area', even though everyone recognizes it's very important. But when you're defending your PhD, you need to defend it in one of our silos.

▶ Can you give me some examples of companies or organizations that do requirements engineering right and perhaps some examples of failures where they haven't done it right?

The companies that claim huge successes are the ones that build critical systems, like the ones in aircraft or engines or cars. But requirements engineering has lots of little successes in areas that are less visible, when consultants help companies understand what they are trying to achieve. They used to be called management consultants or business analysts. But then business analysts got bad press: They are taking your watch, telling you the time, and charging for the privilege!

I think a good requirements engineer would help people write things in precise ways to avoid the imprecision and ambiguities of natural language. They would be able to tell you the technical feasibility of a requirement before you invest too much time moving forwards with the architecture. They exist everywhere, but their job descriptions are not very exciting. I wouldn't even call them requirements engineers, but I would say that they use good requirements engineering practices as part of their daily jobs. When they don't, there are lots of failures in software systems.

For an example, I wrote a paper years ago about the Ariane 5 rocket. It was a short two-pager: 'Ariane 5, Who Done It?' Basically, Ariane 5 was an evolution in the software of Ariane 4, a rocket developed by the European Space Agency to compete with the NASA shuttle. On its first flight, Ariane 5 exploded about 40 seconds after take-off. It turned out that the reason for the failure was a software exception that wasn't handled by the rocket. Instead of getting positioning data, it got error messages, which were used as positioning data. The rocket tried to move to a position that corresponded to that data, which was too extreme. It switched to the backup software, which, because it is software, failed in exactly the same way. Suddenly the rocket found itself in an incorrect position and correctly self-destructed.

When I looked at this, every subdiscipline within software engineering could have argued why it was a problem within their discipline. The testers said not enough testing took place compared to Ariane 4. The designers said that the software was designed as if it were hardware. (You put in redundant hardware to cope with failures, but redundancy doesn't work for software; it just fails in exactly the same way.) I listed all these possible interpretations and a possible requirements interpretation. The requirements for Ariane 5 were different from those for Ariane 4 because the trajectory of the rocket when it took off from the podium was different. It went off at more of an angle than the other.

When they designed the new rocket, the engineers thought that a little bit of software that helps the rocket adjust while it's still on the stand would not be needed, but 'if it's not broke, you don't fix it', so they left it there. This piece of software was triggered and it was thinking, 'Why is this rocket at an angle when it should be completely vertical on the stand?' It went into this whole set of unexpected situations.

The moral is that reusing someone else's requirements can be dangerous. It has some advantages but has some dangers. That was my take on the particular problem.

# This is Simula Research Laboratory

Dedicated to tackling scientific challenges with long-term impact and of genuine importance to real life, Simula offers an environment that emphasises and promotes basic research while still covering the broader landscape from research education to application-driven innovation and commercialisation.

- A non-profit, public utility enterprise organised as a limited company owned by the Norwegian Ministry of Education and Research.
- Funded through basic allowances from the Norwegian Ministry of Education and Research, the Ministry of Trade and Industry, and the Ministry of Transport and Communications, in combination with project funding from the Research Council of Norway, the European Union, the Municipality of Bærum, Statoil, and other partners.
- Conducting basic research aimed at top international level within the fields of communication systems, scientific computing, and software engineering.
- Conducting education and fostering innovation on basis of conducted research.
- Providing a healthy financial basis for the company's operation by actively pursuing public and private funding opportunities to support research, education, and innovation.
- Engaging in business on strictly commercial terms.
- Proud of its international environment and cultural diversity, employing 122 exceptional minds of 30 different nationalities.
- Ranked by the Journal of Systems and Software as the world's most productive institution in systems and software research.
- Host for the Center of Biomedical Computing, which is a Centre of Excellence (SFF) awarded by the Research Council of Norway.
- Host for the Certus Center for Software Verification and Validation, which is a Centre for Research-based Innovation (SFI), awarded by the Research Council of Norway.
- Main research partner in the Center for Cardiological Innovation (SFI), awarded by the Research Council of Norway and hosted by Oslo University Hospital.
- Cooperating with industry to provide solutions and increase research relevance. The largest stand-alone industry collaboration is with Statoil, and is worth 125 million NOK (2005–2015).
- Funded by the Ministry of Transport and Communications for the Resilient Networks Project, which is working to understand the vulnerability of network infrastructure in Norway.
- Since inception in 2001, Simula researchers have supervised 68 PhDs and 249 master's degrees to completion.
- Established in 2001 and headed by Professor Aslak Tveito since 2002.

A. Bruaset, A. Tveito (Eds.), *Conversations about Challenges in Computing*,
DOI 10.1007/978-3-319-00209-5, © Springer International Publishing Switzerland 2013